Health Effects From Hazardous Waste Sites

Julian B. Andelman
Dwight W. Underhill

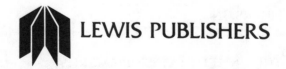 LEWIS PUBLISHERS

Library of Congress Cataloging-in-Publication Data

Health effects from hazardous waste sites.

Includes most of the presentations from the Fourth
Annual Symposium on Environmental Epidemiology, held
at the University of Pittsburgh Graduate School of
Public Health in May of 1983, organized by the
University of Pittsburgh Center for Environmental
Epidemiology, and sponsored by the U.S. Environmental
Protection Agency.
Includes bibliographies and index.
1. Hazardous waste sites—Hygienic aspects—
Evaluation—Congresses. 2. Health risk assessment—
Congresses. 3. Hazardous waste sites—Hygienic aspects—
United States—Case studies—Congresses. I. Andelman,
Julian B. II. Underhill, D. W. (Dwight W.)
III. University of Pittsburgh. Center for Environmental
Epidemiology. IV. United States. Environmental
Protection Agency. V. Symposium on Environmental
Epidemiology (4th : 1983 : University of Pittsburgh
Graduate School of Public Health) [DNLM: 1. Industrial
Waste—adverse effects—congresses. WA 788 H434 1983]
RA567.H45   1987   363.7'28   86-27290
ISBN 0-87371-046-0

Second Printing 1988

LEWIS PUBLISHERS, INC.
121 South Main Street, P.O. Drawer 519, Chelsea, Michigan 48118

PRINTED IN THE UNITED STATES OF AMERICA

Preface

Assessing the adverse human health effects of chemical exposure from waste disposal sites and other point sources is at best difficult. But with the thousands of hazardous waste sites in the United States and abroad, it is necessary that we become more knowledgeable about this ubiquitous problem. This book should provide a useful orientation on the nature, risks, methodologies, and limitations in assessing possible human health exposures and risks from hazardous waste sites.

The five principal parts of this book are:

1. Hazardous Wastes: Scope of the Problem
2. Assessment of Exposure to Hazardous Wastes
3. Determining Human Health Effects
4. Role of Social Groups in Defining Health Risks at Waste Sites
5. Case Studies

The chapter "Evaluating Health Effects at Hazardous Waste Sites," by Gary M. Marsh and Richard J. Caplan, is a comprehensive overview. The remaining chapters are based on presentations at the Fourth Annual Symposium on Environmental Epidemiology, held at the University of Pittsburgh Graduate School of Public Health in May of 1983, and organized by the University of Pittsburgh Center for Environmental Epidemiology, with sponsorship and funding by the U.S. Environmental Protection Agency (EPA) under cooperative agreement number CR-810543. These presentations were edited to make this book, *Health Effects from Hazardous Waste Sites*, appropriate, concise, and complete.

All chapters published in this volume were subject to peer review, consistent with the review policy of the EPA. However, they do not necessarily reflect the views of the EPA or the publisher, and no official endorsement should be inferred. Financial support for the publication of this book has been provided by the EPA under cooperative agreement number CR-812761 with the University of Pittsburgh, and is gratefully acknowledged.

Julian B. Andelman
Dwight W. Underhill

Center for Environmental Epidemiology
Graduate School of Public Health
University of Pittsburgh

Julian B. Andelman is Professor of Water Chemistry at the Graduate School of Public Health, University of Pittsburgh, and Associate Director of the Center for Environmental Epidemiology, an exploratory research center established in cooperation with the U.S. Environmental Protection Agency. He received a BA in biochemical sciences from Harvard College in 1952, and a PhD in physical chemistry from Polytechnic University at Brooklyn, New York in 1960. Following a postdoctoral at New York University, he worked for two years at the Bell Telephone Research Laboratories in electrochemistry. In 1963 he became a faculty member of the University of Pittsburgh, where his research has centered on the chemistry and public health aspects of water supply systems.

Dwight Underhill has a BE in chemical engineering from Yale University and a ScD in industrial hygiene from the Harvard School of Public Health. He is currently an Associate Professor at the Graduate School of Public Health, University of Pittsburgh, where he is also an Assistant Director of the Center for Environmental Epidemiology. For over 25 years, Dr. Underhill has been involved in research in the measurement and control of environmental contaminants, and in this work has served as a consultant to the Nuclear Regulatory Commission, the Frederick Cancer Research Facility, the Lawrence Livermore National Laboratory, the U.S. Army Medical Research and Development Advisory Committee, and the American Society for Testing and Materials.

Contributors

Albert, Roy E., Kettering Laboratory, University of Cincinnati Medical School, 3223 Eden Avenue, Cincinnati, OH 45267

Andelman, Julian B., Department of Industrial Environmental Health Sciences and Center for Environmental Epidemiology, Graduate School of Public Health, University of Pittsburgh, Pittsburgh, PA 15261

Atkins, Patrick R., Aluminum Company of America, 1501 ALCOA Building, Pittsburgh, PA 15219

Caplan, Richard J., Department of Biostatistics and Center for Environmental Epidemiology, Graduate School of Public Health, University of Pittsburgh, Pittsburgh, PA 15261

Clark, C. Scott, Kettering Laboratory, University of Cincinnati Medical School, 3223 Eden Avenue, Cincinnati, OH 45267

Goldman, Lynn R., Children's Hospital Medical Center, Oakland, CA 94609

Harris, Robert H., ENVIRON Corporation, 210 Carnegie Center, Suite 201, Princeton, NJ 08540

Heath, Clark W., Jr., Office of the Center Director, Centers for Disease Control, Atlanta, GA 30333

Kuller, Lewis, Department of Epidemiology and Center for Environmental Epidemiology, Graduate School of Public Health, University of Pittsburgh, Pittsburgh, PA 15261

Magos, Laszlo, Medical Research Council Toxicology Unit, Carshalton, Surrey, SM5 4EF, United Kingdom

Marsh, Gary M., Department of Biostatistics and Center for Environmental Epidemiology, Graduate School of Public Health, University of Pittsburgh, Pittsburgh, PA 15261

Murphy, Patricia, Department of Epidemiology, Graduate School of Public Health, University of Pittsburgh, Pittsburgh, PA 15261

Paigen, Beverly, Children's Hospital Medical Center, Oakland, CA 94609

Papadopulos, Stavros S., S. S. Papadopulos and Associates, Inc., 12250 Rockville Pike, Suite 290, Rockville, MD 20852

Radford, Edward, Department of Epidemiology and Center for Environmental Epidemiology, Graduate School of Public Health, University of Pittsburgh, Pittsburgh, PA 15261

Rodricks, Joseph V., ENVIRON Corporation, 1000 Potomac St, NW, Washington, DC 20007

Schneiderman, Marvin A., Clement Associates, Arlington, VA 22209, Environmental Law Institute, Washington, DC, and Uniformed Services University of the Health Sciences, Bethesda, MD

Schweitzer, Glenn E., Environmental Monitoring Systems Laboratory, Environmental Protection Agency, Las Vegas, NV 89114

Stephens, Robert D., Hazardous Materials Laboratory Section, California Department of Health Services, 2151 Berkeley Way, Room 237, Berkeley, CA 94704

Talbott, Evelyn, Department of Epidemiology and Center for Environmental Epidemiology, Graduate School of Public Health, University of Pittsburgh, Pittsburgh, PA 15261

Tarr, Joel A., Department of Social Sciences, Carnegie-Mellon University, Pittsburgh, PA 15213

Traven, Neal, Department of Epidemiology, Graduate School of Public Health, University of Pittsburgh, Pittsburgh, PA 15261

Williamson, Shelly J., Environmental Monitoring Systems Laboratory, Environmental Protection Agency, Las Vegas, NV 89114

Contents

SECTION I
SCOPE OF THE PROBLEM

SECTION II
ASSESSMENT OF EXPOSURE

SECTION III
DETERMINING HUMAN HEALTH EFFECTS

SECTION IV
DEFINING HEALTH RISKS

SECTION V
CASE STUDIES

Health Effects From Hazardous Waste Sites

SECTION I

Scope of the Problem

Evaluating Health Effects of Exposure at Hazardous Waste Sites: A Review of the State-of-the-Art, with Recommendations for Future Research

Gary M. Marsh and Richard J. Caplan

INTRODUCTION

Every year billions of tons of solid waste are discarded in the United States. These wastes range from common household trash to complex materials in industrial wastes, sewage sludge, agricultural residues, mining refuse, and pathological wastes from hospitals and laboratories.[1] For example, recent estimates indicate that since 1958, the 53 largest chemical manufacturers disposed of 766 million tons of chemical process waste at 3,383 sites in the United States.[2] In 1978 alone these companies generated 66 million tons of chemical process waste. Dump sites tend to be localized in states that have the greatest concentration of industrial activities. Unfortunately, these states are also the most densely populated.

The U.S. Environmental Protection Agency (EPA) estimated that in 1980 at least 57 million metric tons of the nation's total waste load could be classified as hazardous.[1] A hazardous waste is defined by the 1976 Resource Conservation and Recovery Act (RCRA) as ". . . a solid waste or com-

J. B. Andelman and D. W. Underhill (Editors), *Health Effects from Hazardous Waste Sites*
© 1987 Lewis Publishers, Inc., Chelsea, Michigan – Printed in U.S.A.

bination of solid wastes, which because of its quantity, concentration, or physical, chemical or infectious characteristics may cause . . . an increase in mortality or an increase in serious illness . . . or . . . pose a substantial present or potential hazard to human health or the environment when improperly treated, stored, transported, or disposed of, or otherwise mismanaged."[3] In addition, the EPA established four characteristics — ignitability, corrosivity, reactivity, and toxicity — as yardsticks by which to judge the degree of hazard. A list of chemicals falling under this definition was developed, and new substances are added to the list as necessary. To date this list contains over 55,000 chemicals.[4] Estimates by EPA regional offices suggest that there are as many as 50,000 improperly operated sites containing hazardous wastes as defined above.[5]

Persons exposed to toxic chemicals at these dump sites include workers employed in routine operations as well as firefighters, police, and members of special disposal squads who must enter dumps when there are unexpected spills, explosions, or fires.[6] Persons living or working in communities adjacent to dumps are also at risk of exposure, although usually at lower levels. Their exposures may result from inhalation of dusts or fumes dispersed from dumps[7] or from ingestion of wastes which have leached from dumps into the food chain or into ground water used for drinking.[8-9] Hazardous waste sites can contain a large number and variety of chemicals which may produce cancer, disorders of the central nervous system, reproductive alterations, and many other illnesses. The serious potential risk to human health was emphasized within the last few years by events at Love Canal, New York and several other lesser incidents throughout the United States.[10-11]

From a public health standpoint, the contamination of drinking water probably represents the largest waste site-related health problem since it may pose adverse health risks to a large number of people. In particular, EPA has estimated that approximately 90% of the 50,000 improperly operated hazardous waste sites are not environmentally suitable. That is, they are unlined landfill sites or surface impoundments (pits, ponds, lagoons) that will not prevent migration of chemicals.[2] In addition, more than 75% of improperly situated landfill sites are located in wetlands, flood plains, or over major aquifers. By considering these facts along with geologic factors and migration characteristics of chemicals in soils and ground water, EPA has estimated that as many as 30,000 waste sites may pose significant health problems related to ground water contamination.[2] This estimate coupled with the fact that 40% of the U.S. population currently relies on ground water as a source of drinking water suggests that the scope of the health problem that could derive from hazardous waste sites may be enormous.

The potential for hazardous wastes to cause health damage to exposed human populations requires epidemiologic investigations to assess relationships between toxic exposure and possible health consequences, clinical or subclinical. Unfortunately, the classical application of epidemiology is made difficult under myriad methodologic complications and uncertainties related to both exposure and health outcome assessment. For example, investigations may be limited by the small number of persons affected at a particular site, the lack of exposure information, the lack of toxicological data related to mixtures or combinations or chemicals, and the lack of specificity of common clinical indicators of disease. Moreover, health effects of exposure to chemical dump sites depend on the nature of the dumped materials, the integrity of containment, and the underlying hydrogeology and surrounding geography.[12] Much also depends on residential settlement patterns and drinking water treatment.

Health effects of waste site-related exposures may occasionally be acute and overwhelming such as those resulting from seepage of waste arsenic into a household well. More typically, however, exposures are less dramatic and any health effects that exist will most likely be subtle and difficult to detect. They will tend to be chronic and may remain asymptomatic and subclinical for long periods. Moreover, illnesses that do arise from chemical exposure can usually be expected to resemble or be identical with "background" health problems in a community such as nerve damage, cancer, or miscarriage. Also, since many communities surrounding waste sites are small and hard to define, statistical reliability may be inherently impossible to attain.[13]

In short, the United States today is faced with a potentially enormous public health problem that may be difficult to assess with extant epidemiologic methods. This dilemma underscores the need and importance of developing research programs, both basic and applied, that will lead to a more complete understanding and perhaps ultimate solution of the health problems derived from hazardous waste sites.

The main purpose of this chapter is to provide a basis and impetus for future research related to the assessment of human health effects from exposures to hazardous waste sites. In order to achieve this goal the following objectives were established: to describe the extent and magnitude of the hazardous waste site problem from an historical, current, and future perspective; to describe the strengths and weaknesses of various classical epidemiological and statistical approaches for assessing the health impact of human exposures to toxic chemicals found in waste sites; to identify and assess the alternative/nonclassical epidemiological and statistical approaches that are available or may need to be developed for assessing the health impact of waste site exposures; to review current federal and state level activities and research programs in the waste site area; to critically review proposed, ongoing, and published health evaluations; and to appraise the need for and the nature of the research that can provide for more meaningful and valid assessments of health effects from hazardous waste site exposures.

Table 1. Hazardous and Solid Waste Laws

Federal Water Pollution Control Act	1952
Clean Air Act	1963
Solid Waste Disposal Act	1965
Resource Recovery Act	1970
Energy Supply and Environmental Coordination Act	1974
Safe Drinking Water Act	1974
Toxic Substances Control Act (TSCA)	1976
Resource Conservation and Recovery Act (RCRA)	1976
Clean Water Act	1977
Fuel Use Act	1977
Comprehensive Environmental Response, Compensation and Liability Act (CERCLA)	1980

Historical and Current Perspectives

The hazardous waste site problem that exists today is, curiously, an unpredicted effect of the strong sanctions that were enacted during 1952 to 1977 regarding water and air pollution. Because of such legislation as the Federal Water Pollution Control Act (1952), the Clean Air Act (1963), and the Clean Water Act (1977), surface waters and ambient air were no longer considered acceptable sinks for the disposal of wastes and consequently, industries and municipalities turned increasingly to the land for waste disposal.[14] This new trend carried with it such practices as dumping in open land or in unlined earthen lagoons and also resulted in the expansion of the generally more environmentally sound deep well injection and sanitary landfill techniques that had been developed early in the century. While sanitary landfills, in particular, were generally regarded as superior to the open dump, a few articles in the 1950s and a 1961 survey of 250 sites conducted by the American Society of Civil Engineers noted their potential hazards, such as methane fires, ground water pollution, and poor load bearing capabilities.[14] In 1963, in an attempt to generate interest and research in the solid waste area, the U.S. Public Health Service and American Public Works Association sponsored the First National Conference on Solid Waste Research. This conference and others which highlighted the deficiencies in solid waste research were the precipitating force behind the 1965 passage of the Solid Waste Disposal Act. This act created the Office of Solid Wastes, provided the federal government with a more formal role regarding municipal wastes, and paved the way to more comprehensive legislation.

Table 1 shows the evolution of hazardous and solid waste laws that eventually led to the 1980 passage of the Comprehensive Environmental Response, Compensation and Liability Act (CERCLA) commonly referred to as "Superfund."[15]

The 1965 Solid Waste Disposal Act gave the federal government only information and research authority, leaving regulation and enforcement to the states. Amendments added resource recovery as a goal in 1970 but reauthorizations were not noteworthy until the passage of RCRA in 1976. RCRA, the counterpart of CERCLA, provides essentially "cradle-to-grave" regulation of all hazardous wastes and for the first time provided the federal EPA with enforcement power.[3] RCRA does not, however, permit the government to respond directly to the problems caused by improper hazardous waste disposal sites already in existence. CERCLA as the complement to RCRA was enacted in 1980 as a five-year plan to provide federal dollars and authority to respond to emergencies and take remedial action at abandoned hazardous waste dump sites. Costs are to be covered by a $1.6 billion fund, 86% of which is financed by taxes on the manufacture or import of certain chemicals and petroleum and the remainder from general revenues. The fund is reimbursable, and the government generally can take legal action to recover its cleanup costs from those subsequently identified as responsible for the release. The primary responsibility for carrying out the Superfund program has been assigned by Executive Order to EPA. Other federal agencies, such as the Centers for Disease Control in Atlanta, Georgia provide assistance as necessary during a response.

The systematic plan under CERCLA for addressing each site is shown in Figure 1. In general, direct government action, when called for, can involve immediate removals, planned removals, or remedial actions.[16] Remedial actions which are long-term and usually more expensive than the other approaches are aimed at permanent remedies. They may be taken only at sites identified as national priorities. CERCLA required that the 400 worst sites as determined by the site hazard ranking system[17] be addressed first. By the middle of 1983 EPA had inventoried almost 16,000 uncontrolled hazardous waste sites, of which the worst 406 were included on the August 1983 National Priority List (NPL).[18] At that time EPA also proposed 133 new sites in its first NPL update which CERCLA requires at least annually. Figures 2 through 4 show the distribution of the NPL sites by geographic area, type, and route of exposure.

Under CERCLA, state governments, many of which have their own hazardous waste site programs, may plan and manage responses under agreement with the federal government. In remedial actions for which the federal government has lead responsibility, the Army Corps of Engineers manages the design and construction stages for EPA. Private contractors perform the work at a site under federal or state government supervision. The specific guidelines and procedures that the federal government follows when implementing the Superfund law are described in the 1982 regulatory document called the "National Oil and Hazardous Substances Contingency Plan."[17] A later section of this chapter also addresses in further detail these and other federal and state government responsibilities and activities in the hazardous waste site area.

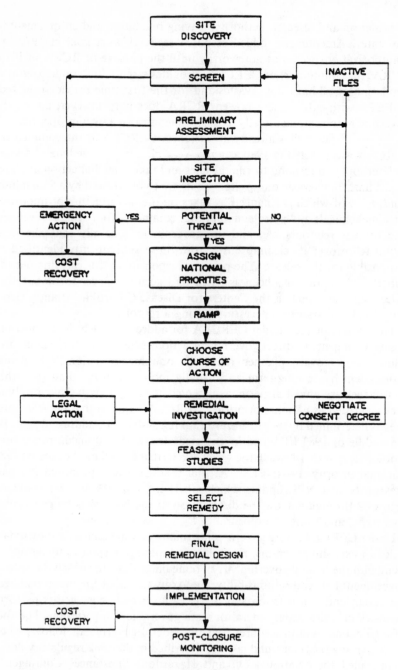

Figure 1.   Superfund site management plan.[16]

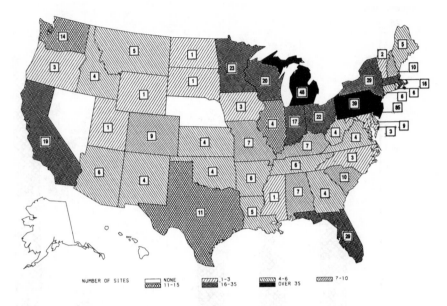

Figure 2. National Priorities List—number of sites per state (totals).[18]

Future Perspectives

It is speculated that site assessments and inspections, remedial action, and other work have been performed at a rate that should successfully address most serious sites and certainly priority sites within the current framework of CERCLA.[16] Nevertheless, the program is not without problems. The sufficiency of funds and the five-year duration of the program may not be adequate to accomplish remediation at all sites that are of public concern. In addition, as pointed out in recently published reports by the Office of Technology Assessment (OTA) and the National Academy of Sciences (NAS), the process of cleaning up Superfund sites may itself lead to the creation of new Superfund sites.[19-20] This concern was based on the observation that materials removed from Superfund sites are usually disposed of in landfills, and the latter have inherent problems no matter how well they are designed. Furthermore, landfills that are now active and managed according to regulations could become future Superfund sites because: (1) no specific compounds are banned from them, (2) many of the materials placed in landfills are highly toxic and remain hazardous for hundreds of years, and (3) nearly all landfills will leak at some time in the future. Regulations under RCRA, however, require monitoring for only 30 years after a landfill is closed. Therefore, disposal of Superfund site materials in landfills unfairly transfers the health risk and part of the cost of waste

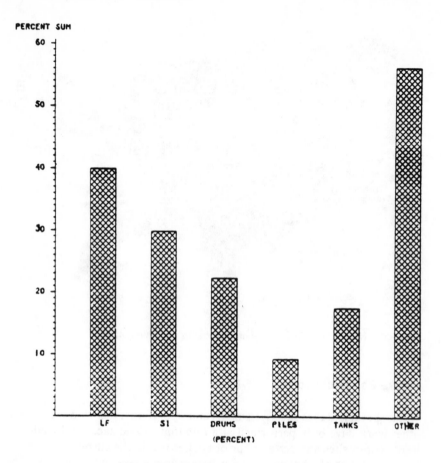

PERCENT SUM

LEGEND· LF - LANDFILLS   SI - SURFACE IMPOUNDMENTS

Figure 3. National Priorities List—types of sites.[18]

disposals to future generations.[21] OTA also noted that in 1985 (the year Superfund initially expires), more sites may need to be cleaned up than are now listed under Superfund as sites requiring attention.[19] Also, the old Superfund sites may not be adequately cleaned up even when they are treated according to regulations. Current laws for cleanup provide no specific technical standards, such as concentration limits, for the extent of hazardous waste removal.

While neither OTA nor NAS believe there is a panacea for all hazardous wastes, they do emphasize the critical need for continued research in the alternative forms of waste disposal (e.g., incineration, thermal methods, recycling, biological degradation) in order to avert the perpetuation of hazardous waste sites.

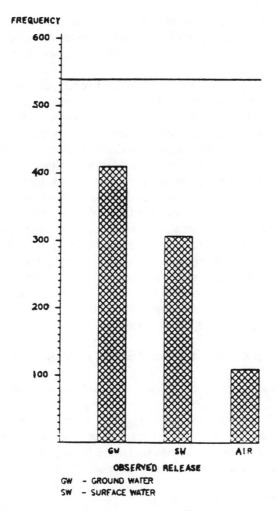

Figure 4.   National Priorities List—kinds of problems.[18]

GENERAL EPIDEMIOLOGIC CONSIDERATIONS IN ASSESSING HEALTH EFFECTS AT HAZARDOUS WASTE SITES

A particularly problematic feature of all health effects evaluations of persons exposed to hazardous waste sites is the sheer diversity of situations in which toxic wastes and human exposures can be involved. No two studies are exactly the same since all waste site situations are unique in some respect. This feature is evident within a subsequent section of this chapter which reviews and summarizes previous, ongoing, and planned epidemiologic investigations of health effects related to hazardous waste site expo-

sures. Such diversity not only prohibits the development of a unified analytic approach to exposure and health outcome assessment but also prevents
the generalization of statistical inferences drawn about a specific waste
site–exposed population.

Regardless of the study design or of the diversity of the underlying setting, however, health effects evaluations of persons exposed to chemical
dumps consist of four fundamental phases: (1) documentation of the nature
and extent of exposure, (2) definition and characterization of exposed and
unexposed populations, (3) diagnosis and measurement of disease and dysfunction in the exposed population, and (4) determination of the relationship between exposure and disease.[22-23] This section presents the general
epidemiologic requirements and related difficulties and limitations associated with these four phases, and the next section proposes and evaluates
specific classical and alternative or nonclassical methodologic approaches
to health evaluation that are derived from the basic epidemiologic tenets. In
these discussions primary consideration will be given to the health effects of
continuous low-dose chemical exposures of a noninfectious nature that
originate from existing common sources (e.g., dump sites, lagoons, and
landfills) as opposed to temporary, intermittent, or episodic exposures
resulting from accidental spills or discharges.

Documentation of the Nature and Extent of Exposure—Phase I

The following sequence of activities characterize this initial phase:

1. *Determine an inventory of the types and amounts of materials present
in the dump through a review of past records, if available, or through
environmental sampling.* While the current technology for analyzing chemicals in samples of air, water, or soil is adequate, there remain the problems
and constraints of deciding where to monitor, deciding what to look for
when the contents of the dump are unknown, overwhelming laboratory
facilities by the need to analyze multiple samples for numerous chemicals,
and monitoring the interlaboratory comparability of sample results.[24]

2. *Determine the toxicity of materials giving consideration to mixtures
and the effect of storage or movement in water or the ground.* The key to
this determination is the availability of toxicity data which, unfortunately,
is unknown or limited for many of the chemicals found in waste sites. These
chemicals are not always end products of familiar commercial processes but
may be the residuals of such processes or even intermediate or precursor
substances. For example, of the nearly 300 chemicals identified at Love
Canal by the New York State Department of Health and the U.S. EPA,
toxicology data were available on fewer than half and for most of these data
were sparse.[25] The National Toxicology Program and other industry pro-

Didn't know seriousness

grams are helping to fill this need for toxicology data; however, their progress has been impeded by the current lack of technology for assessing the chemical toxicity of chemicals in the presence of mixtures and changes that occur on storage or movement in water or ground.[25] The problem of incomplete data is compounded by the absence of a well-developed centralized databank that will make toxicologic data available for chemicals that are likely to be found at waste sites. Efforts are currently underway by the National Library of Medicine and the U.S. EPA Office of Toxic Integration to develop and maintain such systems. In addition, the Chemical Substances Information Network (CSIN) provides a public use record linkage system that accesses data on chemical nomenclature, composition, structure, toxicity, production, uses, health and environmental effects, regulations, and other aspects of materials as they move through society.[26] Appendix A provides more details on the CSIN system.

3. *Determine the nature and quantity of the major environmental emissions from the waste site.* Techniques for assessing the pattern of environmental emissions, a practice referred to by Landrigan[22] as "pollutant mapping" include groundwater and well sampling, soil coring, well drilling, examination of historical operation patterns, and aerial surveys using false-color infrared imagery. Expert consultation may be required from hydrogeologists and from meteorologists to plot the movement of pollutants in water and in air.

4. *Determine the potentiality and probable routes of human exposure.* Modes of exposure may be transcutaneous absorption via direct contact, ingestion of contaminated water or vegetation, or inhalation of contaminated air. Consideration must be given to the size and density of the human populations potentially exposed and their degree of proximity to the toxic materials.

5. *Develop estimates of human exposure.* The most epidemiologically useful and valid estimates are those specific for individual rather than groups, such as breathing zone exposures via personal air samplers or individual water consumption histories.[22] Consideration should be given to the daily dose, time, and duration of exposure of each individual. If possible, external exposure estimates can be confirmed by or supplemented with biological sampling in which a more direct measurement is made of the body burden of a toxic chemical or of the amount excreted. Effective biological sampling requires that consideration be given to the nature of the chemicals involved since some chemicals expose persons only transiently and other chemicals are stored in tissue. Exposure to chemicals which persist in tissue facilitates epidemiologic studies which are undertaken long after waste site-derived exposures have occurred. Well-known examples of

individual biological exposure markers include: levels of pesticides, poly-chlorinated biphenyls (PCB), polybrominated biphenyls (PBB), or other fat soluble chemicals in adipose tissue, serum, or breast milk, levels of arsenic in the urine, and levels of metal in the hair.[27]

There are serious drawbacks related to the practical application of many of the biological markers, however, such as high refusal rates, the differential fat content in males and females, the invasive nature of adipose tissue sampling, and the nonrepresentativeness of lactating women as an exposure group. Other approaches which mitigate many of these problems, such as measuring chemical levels in skin oils, are currently under development.[27]

It must be recognized that regardless of the physical or biological approach taken, some misclassification of exposure status of the individual is likely to occur. As further discussed below, such misclassification of exposure inevitably reduces the power of an epidemiologic study to detect associations between specific exposures and given outcomes. Epidemiologists must resort to constructing less than perfect exposure indices which they then attempt to relate to health phenomena.[28]

6. *Determination of baseline exposure levels in the general population.* The determination of waste site-related exposures is complicated by the fact that most persons, regardless of their residence, possess a body burden of diverse chemicals from the environment.[25] Determining the extent of a waste site exposure requires, therefore, knowledge of baseline levels in the population at large. Currently, the best information available about background levels is for heavy metals, although there remain uncertainties, as with lead, about the minimum levels required to produce health effects in susceptible populations.[29]

The most recent Health and Nutrition Examination Survey (HANES) conducted by the National Center for Health Statistics (NCHS) provided information about body burdens of pesticides and related chemicals, but information about exposures to other chemicals such as PCBs is deficient.[30] Other surveys such as the Human Adipose Tissue Survey conducted by NCHS have revealed significant background quantities of pesticides, PCBs, PBBs, as well as trace amounts of the potent carcinogen, aflatoxin. Moreover, these surveys have demonstrated wide regional variations in observed baseline concentrations.[27]

The limitations and difficulties associated with this initial phase are formidable and not readily rectifiable. For example, without an adequate inventory of waste site materials and their physical storage conditions, it is premature, if not impossible, to design epidemiologic studies. There may be insufficient reason to proceed beyond Phase I if only very low concentrations or small amounts of toxins are present. However, even in the absence of a confirmed toxic exposure, public anger and concern may provide the impetus needed to launch a health evaluation. Unfortunately, even the simplest health effects survey motivated by pressures alone is likely to be unrewarding and possibly misleading since few, if any, measurable health effects are sufficiently specific for exposure to particular toxins or groups of toxins to be surrogates for directly measuring toxic exposure.[23]

Definition and Characterization of Exposed and Unexposed Populations—Phase II

The ability to precisely define and characterize the populations at risk to toxic waste site chemicals depends entirely upon the success achieved in documenting the nature and extent of exposure during Phase I. In particular, of the many Phase I objectives, the ability to develop individual rather than group exposure is probably the most important in terms of accurately identifying high-risk groups and in avoiding dilution effects. These effects are caused by the practice of combining typically small heavily exposed groups with less heavily exposed or unexposed groups in order to increase sample size. In general, these dilution effects translate into a decreased ability or statistical power of detecting a true etiologic association between exposure and health outcome, if one indeed exists.

The choice of an appropriate comparison population should be guided not only by exposure status but also by the primary aim of any epidemiological inquiry—to describe a relationship between exposure and outcome which is, as much as possible, unlikely to be explained by extraneous differences between the exposed and unexposed groups. To accomplish this aim, the study and comparison groups should be at least comparable with respect to known risk factors for the disease(s) under investigation. Examples of such risk factors for many diseases are age, race, sex, ethnicity, occupation, diet, alcohol consumption, and smoking history. As further discussed below, data on each of the potentially confounding or effect modifying variables should be gathered during the investigation so that comparability of the two study groups can be demonstrated and the consistency of association of the potentially confounding variables across various strata or possible interactions between exposure and confounding variables can be assessed.

Measurement of Health Outcomes—Phase III

A simple model of potential adverse health effects caused or potentiated by toxic chemicals from waste sites has been developed by CDC, Center for Environmental Health, as part of their protocol to study PCB exposures at Superfund sites.[31] This model, shown in Figure 5, includes contamination of environmental exposure pathways (e.g., soil, water, food chain) as a necessary condition for human exposure, and human exposure as a necessary condition for potential health effects. A distinction is also made in this model between chemicals with long biologic half-lives (months, years), and

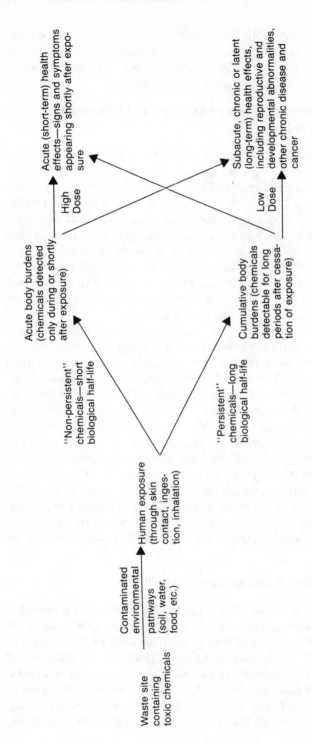

Figure 5. Model of potential human health effects related to chemical exposures from toxic waste sites.[31]

Table 2.   Outline of Health Outcomes for Hazardous Wastesite Studies

I.   Symptoms (rashes, eye irritation)
II.   Signs (rashes, paralysis, tremor, etc.)
III.   Disease or Disorder
    A. Apparent
       1. Abnormal reproductive outcomes
       2. Growth and developmental disorders
       3. Behavioral or psychological disorders
       4. Cancer
       5. Mortality
       6. Other disorders (autoimmune diseases, blood dyscrasias, coronary artery diseases)

    B. Inapparent
       1. Biochemical abnormalities (cholinesterase, erythrocyte protoporphyrin, liver function tests)
       2. Immunologic abnormalities (lymphocyte tests)
       3. Chromosomal abnormalities
       4. Nerve conduction abnormalities
       5. Other test abnormalities (pulmonary function)

chemicals which are rapidly cleared by the body. The body levels of "biologically persistent" chemicals often reflect the lifetime cumulative exposure of the person to these chemicals. Human toxicity, another element implicit in the model, includes not only the immediate effects of producing clinical signs and symptoms, but also the potential for chronic and latent effects such as spontaneous abortions, birth defects, retarded growth and development, liver or renal abnormalities, nervous system and other organ system impairments, and cancer. These and other examples of health effects that could be investigated in epidemiologic studies of waste site-exposed individuals are outlined in Table 2.

In general, the initial selection of possible health effects for study is based on the documentation of potentially toxic exposures conducted in Phase I. The final choice to examine one or several outcomes depends upon the availability of data, validity of outcome measurements, frequency of the outcomes, and biological plausibility. The final choice will also be influenced by the level of effort planned or anticipated for the epidemiologic study. As elucidated below, studies of health effects of waste site exposures can be generally classified according to their underlying level of effort:[33] Level I (based on existing accessible exposure and outcome data), Level II (involves collection of more precise exposure and outcome data, as well as data on confounding variables), and Level III (involves prospective followup of exposed and unexposed cohorts). For example, Level I studies may draw on vital certificate data or special registries of tumors or malformations, in order to examine birth weights, perinatal mortality, cancer incidence or mortality, or sex of offspring, whereas Level II and Level III studies can include outcomes identified in medical records (spontaneous abortions, malformations, behavioral or psychological disorders), through interviews with study subjects (spontaneous abortion, sexual dysfunction, symptoms or signs of rashes, paralysis, eye irritation, etc.), or through biological studies of study subjects (biochemical, immunologic, and chromosomal assessments, and nerve conduction velocities).

Of course, the converse of the above general situation may occur. That is, the predetermined choices of health outcomes may greatly influence the level of effort and the specific type of study that might be conducted. For example, relatively acute effects such as subclinical liver dysfunction or dermatitis might be best addressed in a clinical epidemiologic study of morbidity where certain tests and examinations might be administered to the entire sample of subjects selected for study. Cancers of internal organs, usually associated with long latency periods (measured from first exposure to incidence or death) would be best explored in a longitudinal mortality study or by establishing a registry. Certain reproductive outcomes such as birth defects might be best studied by developing a registry, while certain other outcomes such as low birth weight could be measured by retrospective analysis of birth records in community hospitals.[34]

Another important consideration is that the potential effects to be examined in children may be different from those sought in adults. In fact much of the existing chemical toxicity data has been gleaned from studies of adults and animals; relatively little is known about the effects on children.[34] Specific health outcomes and study designs that may be incorporated in waste site evaluations are discussed in detail below in this and a subsequent section.

The number of end points to be examined is another important issue that requires careful consideration in health effects evaluations. Even in ideal situations where waste site exposures and their related toxicity are well known, it is not necessarily the best approach to attempt to measure all possible health outcomes related to the exposures.[34] There are several reasons for limiting the number of health outcomes for study. In general, due to the complexities of testing large numbers of people, it is extremely difficult in a community survey to examine a large number of variables in a standardized fashion. A better approach is to examine a few highly relevant variables utilizing well-standardized procedures and survey methods that are validated and relatively easily performed in a community survey setting. Furthermore, while there may be multiple health effects associated with exposure to a particular substance, the most sensitive indicator affected should be chosen for measurement under ideal circumstances.[34] The statistical ramifications of limiting the number of health outcomes for study is discussed below.

The above discussions apply to the relatively ideal situation in which health outcomes to be studied are determined with prior knowledge of the nature and extent of toxic exposures. In many cases a different situation occurs when it is not possible to document exposures directly, and it may be necessary to infer the exposures indirectly from the health effects themselves. In this situation, the signs and symptoms must be first meticulously analyzed so that a clinical syndrome can be accurately defined. Next possible etiologies of this syndrome must be considered, usually including both toxic and nontoxic causes of the disease. When a suspect chemical or group of chemicals is identified, epidemiologic techniques, tissue analyses, and animal experiments can provide further information.[35]

Finally, a third waste site situation could possibly arise in which no known toxic exposures are documented and the population displays no signs or symptoms of disease or dysfunction. In this situation, it is questionable whether the epidemiologist or other health specialist has any role to play. There is a high risk of totally fruitless investigation or, even worse, of positive preliminary findings that eventually prove spurious. For these reasons it is often argued that researchers in this situation should avoid "fishing expeditions."[32,35]

In recent years at least four large meetings were convened to consider the public health and technical problems associated with hazardous waste sites and to evaluate approaches to both monitoring exposures to toxic substances and to linking these exposures to possible adverse health effects. Proceedings of the three meetings held in 1981 have recently been published.[36-38] The proceedings of the fourth meeting (University of Pittsburgh, Fourth Annual Symposium on Environmental Epidemiology) appear in this volume. Of the many exposure/health indices considered at these meetings, the most attention was placed on three major areas believed to be the most promising for epidemiologic investigation: reproductive alterations, chromosomal damage, and neurotoxic effects. Although no conclusive answers were produced, the general consensus of the meetings seems to be that while these abnormalities can provide approaches to monitor chemical exposure, it remains extremely difficult to relate that exposure to long-term health effects. The remaining part of this section summarizes the relative advantages and disadvantages of these and other potentially useful end points that can be measured to examine the potential toxic effects of chemicals in human populations.

Reproductive Effects

The June 1–2, 1981 Rockefeller University Symposium concluded that reproductive dysfunctions are the most promising end points for epidemiologic investigations of waste site-exposed populations.[13] Table 3 abstracted from Warburton[39] provides a partial list of the analytic indices of reproductive dysfunction that may serve as end points in health studies. Sources of reproductive data include: vital statistics records (such as birth and death certificates, amniocentesis and newborn blood sampling records, and chro-

Table 3.   Analytic Indices of Reproductive Dysfunction

Sexual Dysfunction
    Decreased libido
    Impotence

Sperm Abnormalities
    Decreased number
    Decreased motility
    Abnormal morphology

Subfecundity
    Abnormal gonads, ducts, or external genitalia
    Abnormal pubertal development
    Infertility (of male or female origin)
    Amenorrhea
    Anovulatory cycles

Illness During Pregnancy and Parturition
    Toxemia
    Hemorrhage

Early Fetal Loss (to 28 weeks)

Perinatal Death
    Late fetal loss (after 28 weeks) and stillbirth
    Intrapartum death
    Death in first week

Decreased Birth Weight

Gestational Age at Delivery
    Prematurity
    Postmaturity

Altered Sex Ratio

Multiple Births

Birth Defects
    Major
    Minor

Chromosome Abnormalities (detected in early fetuses, through amniocentesis, in
    perinatal deaths, in livebirths)

Infant Mortality

Childhood Morbidity

Childhood Malignancies

Age at Menopause

mosome and malformation registries); hospital and physician records; and interview data. The following discussion is limited to five of the most commonly studied outcomes: infertility, sperm abnormalities, spontaneous abortion, congenital malformations, and low birth weight.

Impotence and sterility or infertility are generally not easily studied measures since they are difficult end points to measure accurately, and there are many confounding variables in coital and contraceptive practices and in the unrecognized or unreported loss of fetuses.[39] Only when the health effects are striking, as was the case with kepone workers in Virginia[40] and dibromochloropropane workers in California[41], do these measures become worthwhile indicators.

Semen abnormalities can be analyzed as indicators of exposure; however, their significance for reproductive function is largely unknown. Even among fertile men sperm numbers and ejaculate volumes vary greatly, as may sperm motility, i.e., swim rate. Single samples are insufficiently representative, and, in populations under study, as many as half the men asked will refuse to provide specimens. Sperm morphology, i.e., structure, can also be useful if the frequency of abnormalities is sufficiently unusual, although it is subject to large interobserver variability even in the presence of standardized classification schemes.[33] However, abnormal morphology has been shown to be a more sensitive measure than sperm counts, since, for example, a 25% change in the frequency of abnormalities can be detected with only 26 men, whereas a similar change in sperm count requires 200 men for detection.[42] Morphological studies have been useful in certain cases such as in a study of carbaryl pesticide production workers where sperm counts were normal, but there was a siginificant increase in the number of sperm abnormalities.[43] Overall, the Rockefeller University Symposium concluded that "except in extremes, moderate variation or abnormality of sperm count, motility, or morphology cannot be correlated with reproductive dysfunction or be used as strong proof of health damage in a population."[13]

Spontaneous abortion, defined as termination of pregnancy before 28 weeks of gestation, may be a good indicator of reproductive dysfunction and exposure to toxic substances.[39] Normally, spontaneous abortion acts to screen out about 90% of all chromosomally abnormal conceptions and about 50% of all malformed fetuses.[39] At least 15% of all recognized pregnancies are spontaneously lost, and the number in unrecognized pregnancies may be much higher. This high rate, like that of sperm abnormalities, makes spontaneous abortions one of the outcomes having the greatest power to detect a change given a limited population.[33] That is, for example, a doubling in the rate of occurrence of a rare event is much more likely to occur by chance than a smaller increase in the rate of a more common event. Spontaneous abortions also have the advantage of providing a specimen that can be examined in various ways to provide information about the mechanisms of loss, including chromosomal analysis and detailed morphological classification of defects.[39]

There are several disadvantages to studying spontaneous abortions. First, the only sources of available data are medical records and personal interviews. Medical records are insufficient because a large number of spontaneous abortions are not brought to medical attention, and interview data are subject to recall bias and to inaccuracies in self-diagnosis. Secondly, chromosomal and pathological examination of the abortus is also expensive.[33]

Increases in spontaneous abortion rates have been reported to occur among female anesthesiologists in England and the United States,[44-46] in the wives of vinyl chloride workers,[47] and in women near Alsea, Oregon and at Love Canal exposed to the herbicide 2,4,5-T.[48-49] However, as described by Warburton,[39] it is believed that each of these studies has serious deficiencies which make the data suspect.

Congenital malformations or birth defects are much less common, occurring in only 2 to 3% of live births. However, as the most distressing and visible outcome, they are most likely to be self-reported by the community as having increased.[39] Anomalous fetal development can have a great many causes, both genetic and nongenetic, and can take a wide variety of physical manifestations. Maternal drug ingestion (thalidomide, anticonvulsants) during pregnancy has been demonstrated to lead to increased rates of specific malformations, and some studies have suggested increases due to environmental exposures such as anesthetic gases, lead, or 2,4,5-T.[39]

However, in the case of chemical waste sites, where multiple agents are involved, the effects could be so varied that it could be almost impossible to demonstrate any linkage. Until very recently no systematic records of malformations were kept, and there was little agreement as to their classification. Several new registries are being developed that may permit future analyses of birth defect patterns.

Low birth weight is a useful indicator of maternal health, and is strongly associated with neonatal death, congenital malformations, developmental delay, and chromosomal abnormalities.[39] About 7% of American infants weigh less than 2500 grams at birth. One of its chief advantages as a diagnostic tool is that birth weights are routinely collected on birth certificates and hospital records in most parts of the world, and the data are not subject to biases of reporting.[33,39] Birth weight also has considerable power in statistical analysis due to its relatively high frequency and because it is a continuous variable. For these reasons, the reported increase in low birth weight infants is the most convincing reproductive effect described from Love Canal[49] and from the areas surrounding lead smelters in Sweden.[50] In comparing birth weights between two groups, it is important to account for several extraneous factors known to be strongly associated with low birth weight: maternal age, parity, race, smoking habits, and gestational age.[39]

Cytogenetic Effects

Cytogenetic analyses provide ways of directly visualizing human chromosome damage and estimating genetic effects. From distinctions of size and shape, it is possible to classify the 46 human chromosomes and to identify breaks and other anomalies. As elucidated by Wolff,[51-52] effects can show up as chromosome aberrations or as *sister chromatid exchanges* (SCE). Such observations are most commonly made in white blood cells because these are easy to obtain. Compared to chromosome aberrations, SCEs are a more sensitive indicator for exposure, often showing significant increases in number at chemical doses a hundredth as large as those necessary to increase the yield of aberrations.[51-52]

Cytogenetic methods have proven useful for assessing chromosome damage from radiation as in the case of a recent study which examined chromosome aberrations in a group of atomic submarine refuelers in Britain.[53] It was judged at the Rockefeller symposium[13] that the utility of these methods for assessing chromosome damage from chemical exposures is limited, however, since: (1) the chemical dose within the cell nucleus cannot be determined; (2) the rate of chromatid or chromosome repair is unknown; (3) with chemical mutagens, chromatid exchange is much more likely than whole chromosome damage, the background incidence of chromatic exchange is higher than that for chromosome exchange, and the mechanisms that repair chemically induced breaks appear to be more efficient than those after radiation damage; (4) chromosome aberrations that are so severe as to be detectable tend to lead to cell death, so they are not passed on to succeeding generations; (5) there is wide variability among people, and thus reference populations, in SCE responsiveness; and (6) chromosome abnormalities can be caused by many ubiquitous agents such as medical and dental X-rays, caffeine, viruses, and other environmental exposures making specific chemical exposure assessment and the definition of reference populations very difficult.[13,33,51,52] Furthermore, since SCEs in particular do not appear to affect the health of cells, they do not allow any conclusions about human health damage, although they can serve as a useful index of chemical exposure.[51-52]

In attempts to monitor people exposed to suspect toxic chemicals, investigators have found only a very few compounds that elevate the mean number of SCEs above that of reference populations.[51-52] Studies showing increased SCEs include laboratory workers who carry out hormone analyses and organic chemical research,[54] persons handling cytostatic compounds,[55] and hospital workers exposed to the sterilant ethylene oxide.[56] These findings were somewhat confounded by cigarette smoking patterns, however, since cigarette smoking itself can cause SCEs.[51-52]

The Rockefeller symposium concluded that ". . . cytogenetic assays hold promise as an index to chemical exposure but they do not allow firm conclusions about health damage either to individuals on whom they are performed or to the populations of which those people are members."[28]

Neurotoxic Effects

At the National Institute of Environmental Health Sciences and Rockefeller symposia, Schaumberg et al. review and critique current methods for monitoring potential neurotoxic effects of exposures to materials from hazardous waste sites.[57-58] The authors note that neurotoxic illness is generally not focal but diffuse; degeneration may occur in neurons, in their myelin sheaths, or in axons, i.e., nerve junctions, throughout the body. Neurotoxic impairment may thus produce a wide variety of clinical symptoms. Moreover, subclinical impairment of the nervous system is prevalent and generally unnoticed throughout the population, and further impairment is not recognized unless it is substantial.

Certain general neurotoxicological properties are conducive to the study of hazardous waste site exposures. Neurotoxins usually seem to have thresholds for onset of toxicity, making it possible to conceive of no-effect levels. Chemicals tend to exert their effects on the nervous system immediately rather than after a latent period. Effects can differ dramatically depending on dose.[57-58] For example, massive exposure to acrylamide causes convulsions, hallucinations, and encephalopathy, whereas milder exposure may cause peripheral nerve damage and blindness.[59]

Nerve conduction velocity (NCV) and evoked-potential studies are among the most widely employed electrophysiological approaches to detect peripheral and central nervous system dysfunction. Measurement of NCVs is simple and noninvasive, can be performed with portable equipment, and is widely employed in field assessment of polyneuropathy. The utility of NCV studies is limited, however, since they are not sensitive to minimal dysfunction and their valid use requires considerable expertise, mature judgment, and careful standardization.[57-58] A Danish report of lead-exposed workers serves as a model of a meticulous NCV study.[60] Recent advances in computer averaged evoked potentials enable the noninvasive recording of synchronous activity from its source within the nervous system. The possible effects of n-hexane[61] and xylene and alcohol[62] on the human visual system, and those of lead[63] on the somatosensory system have been investigated in humans using the evoked potential method. The use of evoked-potentials as a neurological screening procedure in waste site studies appears limited, however, both by the variability of the measure when applied to populations at large and the current lack of standard recording procedures among laboratories.[57-58]

Other screening devices are currently being developed which are targeted at specific functions vulnerable to the neurotoxin in question. For example, the fingertip sensation test which utilizes a modified Optacon® instrument (converts visual images into tactile forms) shows promise as a simple reliable technique for detecting sensory loss and abnormal metabolic or toxic conditions.[64-65]

Schaumberg et al.[57-58] also note that the new noninvasive imaging procedures (positron emission tomography [PET] and nuclear magnetic resonance [NMR]), which monitor metabolic activity in areas of the brain, may provide insight into the effects of neurotoxins. Currently, however, these techniques are expensive, require experienced observers and cannot be used in the field.

A wide range of neuropsychological deficits has been reported in association with toxicants including: disturbances in overall intelligence, memory, and problem solving, and alterations in attention level, psychomotor functioning, sleep behavior, and mood.[66-67] Unfortunately, detection of these effects in large, previously unstudied populations is difficult.[57-58] Clinical evaluations are highly subjective and symptomatology is usually vague. Additional problems are absence of standardization and nonspecificity of many of the available tests. Schaumberg et al. note that the recent study of memory performance in polybrominated biphenyl (PBB)-exposed individuals[68] exemplifies a well-constructed evaluation of neuropsychological status and should serve as a model for future work in this area.

In general, there currently exists skepticism about the ". . . applicability of neurotoxicological diagnosis to persons who live with very low chronic exposures to mixtures of chemicals, but who have no medical symptoms of illness, as is typically the case with populations residing near waste sites."[57]

Effects on Tissues, Organ Function, and Biochemical Markers

In addition to the blood, from which white blood cells can be analyzed for chromosomal damage, tissues from various organs such as liver, kidney, bone marrow, and placenta can be examined for enzymatic and morphologic discrepancies. With the exception of placenta obtained at birth, however, these examinations require invasive procedures and are not suitable for apparently well individuals.

Aside from the assessment of neurotoxic effects described above, there are a number of tests for organ function such as of the liver, kidneys, heart, and lungs which can be used to detect damage from toxic chemical exposures. These tests are feasible from the standpoint of the collection of samples such as blood or urine on which to perform the tests. Some of the difficulties, however, are that "normal" values are rather broad, often depending on age, sex, and other characteristics of the individual tested, they are generally less discriminating for disease, and often many factors other than exposure to toxic agents influence results.[2]

For example, most liver function tests are done on patients with a variety of illnesses rather than on normal subjects, and clinical laboratory tests are not sufficiently standardized to detect small differences.[69] Liver function tests were shown to be useful indicators of exposure and dysfunction, however, in a study population residing near a large toxic waste dump located in Hardeman County, Tennessee in which area residents were exposed to toxicants via contaminated drinking water. A comprehensive evaluation of the exposed population revealed evidence of hepatomegaly and elevated liver function tests apparently caused by ingestion of numerous organic chemicals, including several known hepatotoxins.[70]

Plasma lipid values have been established in lipid research clinic population studies,[71] and there are a variety of other biological markers which show potential for detecting minimal disease or very early disease among persons exposed to toxic chemicals. These include various types of stress tests (as used to assess exposures to carbon monoxide), immunologic markers (e.g., autoantibacillus, and thyroid, pancreatic, and liver antibodies), and noninvasive or minimally invasive measurements (e.g., CAT scanning and NMR).[72] Again, however, the basic problem with these approaches is the uncertainty of whether the measurements are directly related to disease.

Cancer Incidence

In the context of toxic chemical exposures, the occurrence of human cancer is frequently viewed as the result of somatic mutation and thus a potential indicator of genetic damage.[2] Its feasibility for application in the assessment of chemical waste site damage is limited, however, by its infrequency and a prolonged latency period between exposure and disease, usually measured in years or decades. Moreover, as with morbidity data in general, the interpretation of cancer incidence data is made difficult due to changes in ascertainment or in definitions of disease and to a lack of background incidence data in sufficient geographic and demographic detail. For example, cancer incidence data are available at the county or local level only in those states which have active tumor registry programs, or in those standard metropolitan statistical areas (SMSA) included in NCI's Surveillance, Epidemiology, and End Results (SEER) Program. (A subsequent section in this chapter provides further discussion of these and other databases).

The study of cancer incidence as a health outcome in a waste site exposure situation will require, at a minimum, utilization of an epidemiologic study design that accommodates the rarity and long latency periods associated with most cancers. Several such designs and examples are described in the next major section.

Mortality

The utility of mortality as a health outcome in hazardous waste site evaluations is limited primarily because it is of limited value in studying rare diseases, and mortality data are not generally available for small geographic areas. Moreover, mortality data are derived from death certificates which are often incomplete and may not accurately reflect the true underlying cause of death.

As with cancer incidence, the study of mortality from relevant chronic diseases in the waste site exposure setting will require the use of special study designs, such as the prospective or historical-prospective followup study. These designs rely heavily, however, upon the availability of national and state level databases, such as the National Death Index, that can be used to determine the vital status of study subjects. (Examples of this approach in waste site studies are provided below.)

In order to better utilize mortality data in waste site studies, the National Center for Health Statistics (NCHS) should be urged to generate more current and geographically specific data on total and cause-specific mortality. Also, emphasis should be placed on the sentinels of disease—those diseases that are relatively rare and that may be associated with environmental exposures (e.g., osteogenic sarcomas, rhabdomyosarcomas) and those cancers and other diseases that occur in relatively young age groups.[72]

Determination of the Relationship between Exposure and Health Outcome—Phase IV

The ultimate objective in epidemiologic studies of persons exposed to hazardous waste site materials is to associate particular exposures with potential biologic effects and thus define cause-effect relationships. Such associations are considerably strengthened if dose-response relationships can be found, that is, if increasing levels of exposure are associated with increasing frequency of the biologic effect.[23] The achievement of this objective is made difficult, however, not only by the limitations which are inherent in all observational studies of human populations but also by the number of particularly complex real-life situations which uniquely characterize studies of hazardous waste site exposures. In this context, the fullest exploration of human observational studies is often greatly restricted by concern for confidentiality on the part of exposed and affected persons, for parsimony by health authorities, for safeguards on the part of industry, and for political considerations on the part of government agencies.[28]

More specifically, epidemiologic studies of populations exposed to toxic waste site materials are likely to be limited by the following technical and human problems:[28,33]

1. Populations living in the vicinity of a hazardous waste site are usually small thus limiting both the range of outcomes and the size of the effects that can be studied.
2. Persons living in any given area are usually heterogeneous either with respect to characteristics that can influence many health outcomes independently of exposure (age, race, socioeconomic status, occupation, smoking, alcohol consumption) or with respect to the type, level, duration, or timing of exposure. Moreover, there is in- and out-migration and geographic mobility within areas.
3. Actual population exposures are generally poorly defined.
4. Many of the health end points of interest are either rare (e.g., specific malformations), are associated with long or variable latency periods (e.g., cancer), or are unlikely to have been routinely recorded prior to the investi-

gation (e.g., spontaneous abortions). In addition the instruments used to measure health outcomes (e.g., questionnaires) are generally very insensitive.

5. Publicity related to the episode under study may produce or accentuate reporting bias.

6. The conduct of waste site studies is made difficult due to the presence of a highly charged atmosphere of anger and fear which often accompanies suspicion of adverse health effects.

The specific requirements and limitations associated with the assessment of exposures, the definition of exposed and unexposed populations, and the choice of health outcome (Phases I–III) have been discussed in the previous three sections. This section now discusses the aforementioned methodologic problems and how they generally affect the statistical aspects of any well-designed study of toxic exposures, in particular, statistical power, bias, and interaction. How these methodologic limitations affect the choice, conduct, and statistical aspects of specific epidemiologic study designs is discussed in the next major section.

Statistical Power

In the context of hazardous waste site studies, statistical power can be defined as the probability that an adverse health effect of a specific size will be detected when it is present in the target population from which the sample was drawn. Power is an extremely important consideration since it helps to determine study design and provides an objective basis from which to meaningfully interpret study results. Statistical power is a function of the following study parameters:

1. *The size of the study and control groups.* In general, power increases as sample size increases.

2. *The variability of the health outcome under study.* For discrete events this will depend on the usual or expected rate of the event in the control population. In general, power is inversely related to the variability of the health outcome in the target population.

3. *The predetermined statistical significance level or type 1 error that will be accepted as confirmation of an association between exposure and health outcome.* This assumes a specific probability model of which more than one may be feasible. With all other parameters fixed, power is directly related to the significance level. Besides the choice of the significance level for a specific test, the probability of falsely concluding that an association is present at least once is affected by the number of associations that is tested within a single analysis. This phenomenon, referred to generally as the "multiple comparison problem," arises frequently in epidemiologic research. Stated more precisely, the probability of falsely claiming statistical significance at a level α in at least one of n independent comparisons is $1 - (1 - \alpha)^n$. Table 4 shows this value, known as the experiment-wise error

Table 4.   Probability of Falsely Claiming Statistical Significance [1 − (1−α)^n] at a Level α in at Least One of n Independent Comparisons

n	α = .05 1−.95^n	α = .01 1−.99^n
1	.0500	.0100
2	.0975	.0199
5	.2262	.0490
10	.4013	.0956
25	.7226	.2222
50	.9231	.3949
100	.9941	.6339

rate, for selected values of α and n. This table reveals that the cumulative probability of making at least one type 1 error can be substantially larger than when multiple hypotheses are tested.

The experiment-wise error rate can be controlled statistically via various simultaneous inferential procedures,[73] however, the net effect is usually a decrease in overall power. Alternatively, the multiple comparisons problem is probably better controlled in waste site studies by limiting the number of exposure variables and/or health outcomes in the inferential analysis.

4. *The magnitude of the expected association between exposure and outcome.* With all other parameters fixed, power is directly related to this magnitude.

5. *The design of the study and statistical techniques used for analysis.* There are several special design and analytic techniques that may be used to enhance power. These include: refining the history of exposure to avoid misclassification bias; refining the response variable to conform with an anticipated biologically coherent health outcome; increasing sample size via intensified case finding; forming composite exposure or outcome variables; use of continuous rather than discrete health outcome variables; use of repeated measures on each study member; stratification or matching; and clustering techniques.[28,33]

The interrelationships of the primary study parameters that determine statistical power are illustrated in Figures 6 and 7. Figure 6, which pertains to cohort or cross-sectional studies, shows the relationships between health outcome frequency, sample size (in study and control group), and the magnitude of the effect that can be demonstrated at the two-tailed 5% significance level with a power of 80%.[74] For example, to detect at these statistical error levels an increase in illness prevalence from 0.1% to 0.2% would necessitate about 25,500 unexposed and 25,500 exposed persons; whereas a tenfold increase over the same background rate (0.1%-1.0%) would require only about 1000 persons in each group. While the smaller increase in prevalence hypothesized above may not seem significant, its extrapolation to the total U. S. population (226 million) would affect an additional 226,000

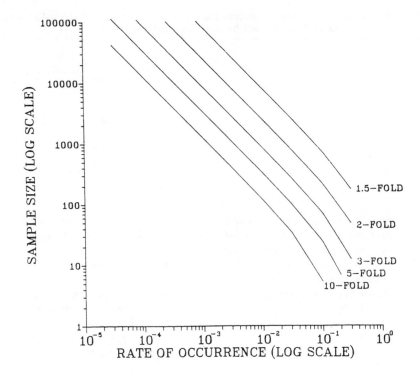

Figure 6. Rate of occurrence vs. required sample size to detect an n-fold increase in rate. $\alpha = .05$ (2-tailed), power = .80.

people.[28] In general, Figure 6 shows that for a given sample size the power to discern modest effects increases with increasing frequency of the event under study, or for a given level of frequency the ability to detect a given effect size increases with increasing sample size. It should be noted that power in cohort studies can be enhanced by increasing the control group size relative to the size of the study group.[74]

In a similar fashion, Figure 7 shows for unmatched case-control studies the relationship between the proportion of controls exposed, sample size (in case and control group), and the minimum relative risk that can be detected at the two-tailed 5% significance level with a power of 80%.[75] For example, in a community where 25% of the unaffected population is exposed to a toxic waste site material, an unmatched case-control study conducted at these statistical error levels would require about 150 cases and 150 controls to demonstrate a twofold increase in risk; whereas a tenfold increase in risk could be detected with only about 15 cases and 15 controls. In general, Figure 7 shows that for a given case and control group size, the power to discern modest effects is maximized when the proportion of exposed con-

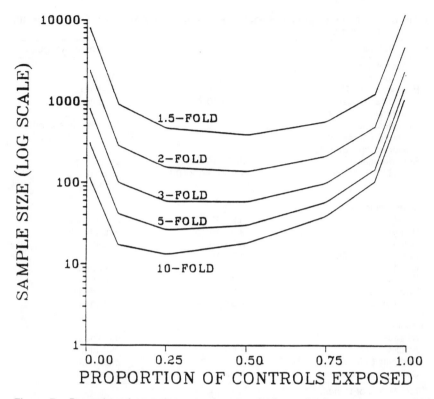

Figure 7. Proportion of controls exposed vs. required sample size to detect an n-fold relative risk. α = .05 (2-tailed), power = .80.

trols is around 0.25 to 0.50. Conversely, for a given proportion of exposed controls, the ability to detect a given effect size increases with increasing sample size. It should be noted that power in case-control studies can be enhanced by individual matching and/or by selecting multiple controls for each case.

Although the above illustrative examples demonstrate that large (i.e. ten-fold) increases in background disease frequency could be detected at acceptable error levels in cohort or case-control studies with quite manageable sample sizes, it is unlikely that the relatively low levels of toxic chemical exposures that prevail in most waste site situations would produce excesses in disease of this magnitude.

Finally, in order to place Figures 6 and 7 in proper perspective, Table 5 provides the frequencies of occurrence of selected health outcomes that were alluded to above.

Table 5.   General Background Frequencies and Units of Analysis of Selected Health Outcomes

Health Outcome	Frequency	Unit of Analysis
Reproductive Effects[a]		
Azoospermia	1×10^{-2}	Males
Birth weight <2500 g	7×10^{-2}	Livebirths
Spontaneous abortion after 8–28 weeks of gestation	$1-2 \times 10^{-1}$	Pregnancies
Chromosomal anomaly among spontaneously aborted conceptions	$3-4 \times 10^{-1}$	Spontaneous abortions
Birth defects	$2-3 \times 10^{-2}$	Livebirths
Neural tube defects	$1 \times 10^{-4}-1 \times 10^{-2}$	Livebirths and stillbirths
Cancer Incidence[b]		
All sites	3.2×10^{-3}	Individuals
Stomach	9.8×10^{-5}	Individuals
Colon	3.3×10^{-4}	Individuals
Lung and bronchus	4.5×10^{-4}	Individuals
Bladder	1.5×10^{-4}	Individuals
Kidney	6.4×10^{-5}	Individuals
Lymphomas	1.2×10^{-4}	Individuals
Leukemias	9.5×10^{-5}	Individuals
Mortality		
All causes	9.5×10^{-3}	Individuals
All cancer sites	1.6×10^{-3}	Individuals
Cirrhosis of liver	1.5×10^{-4}	Individuals
Congenital anomalies	8.4×10^{-5}	Individuals

[a]From March of Dimes, 1981.[33]
[b]Average annual age adjusted (1970) incidence rates, 1973–76, all SEER sites.[114]
[c]1970 U.S. mortality rates.

Bias

Since the ultimate aim of any study is to describe an exposure-outcome relationship that is unlikely to be explained by extraneous differences between the two study groups, it is imperative that two sources of variation be controlled: variation in the characteristics of the study groups that relate to the a priori chance of exposure or to outcome, and variation in the quality of data collected for the two study groups.[28] The inability to control for these sources of variation can lead to a biased estimate of the exposure-outcome relationship.

These and other sources of bias resulting from methodological features of study design and analysis can be classified in a variety of ways, such as the catalog of biases provided by Sackett.[76] Using for the most part a terminology for classification developed by Miettinen,[77-78] the following discussion describes the primary types of biases that are likely to be encountered in epidemiologic studies of waste site-exposed populations.

Confounding bias. Kleinbaum et al.[78] provide a general definition of a confounder as "a risk factor for the disease (outcome) under study whose control in some appropriate way (either singly or in conjunction with other variables) will reduce or completely correct a bias when estimating the true exposure-disease relationship." The basic notion of confounding is best explained, however, in the context of the two general categories of epidemiologic study designs, the retrospective (case-control) and prospective (cohort).[28] In a case-control study, case and control populations should be comparable in their a priori chances of exposure and in characteristics other than exposure which relate both to the outcome under study and to the exposure variable. Likewise in a cohort study, exposed and unexposed groups should be comparable in all known risk factors for the outcome. As discussed above, individual characteristics such as age, race, sex, cigarette smoking, diet, alcohol consumption, and occupation are known to be related to many of the health outcomes that might be examined in waste site studies. If these characteristics are also related to exposure status then they represent "confounders" and must be controlled either through design and/or statistical analysis in order to prevent confounding bias. Moreover, even if these characteristics are not related to exposure, statistical control may remain useful by removing unwanted sources of variation and increasing the precision of the estimate of association.[28] Finally, it should be noted that if certain individual characteristics are related to exposure but not to health outcome then their statistical control should be avoided since it will lead to decreased precision in the estimate of association.

Kleinbaum et al.,[78] Breslow and Day,[79] Schlesselman[75] and Anderson et al.[80] serve as good references to design-analytic approaches for detecting and controlling confounding bias in epidemiologic research. In toxic waste site studies, however, it is likely that the control of confounding bias will be especially limited by practical considerations, such as the inability to collect from the two study groups sufficiently detailed data on potential confounders (e.g., smoking or dietary histories).

Selection bias. This term refers to a distortion in the estimate of effect resulting from the manner in which subjects are selected for the study population.[78] Sources of selection bias include: flaws in the study design, most notably concerning the choice of groups to be compared (all types of studies); the choice of the sampling frame (particularly case-control and cross-sectional studies which are discussed in the following section); loss to followup or nonresponse during data collection (in cohort studies); and selective survival (in case-control and cross-sectional studies). Selection bias can also result in case-control studies when the procedure used to identify disease status varies with exposure status. For example, in cohort studies, those in the exposed group who become ill may be diagnosed and treated earlier than those in the control group. Selection bias can be reduced or eliminated via continued followup or periodic and comparable examinations of both the exposed and control groups.[69]

Information bias. This bias, also referred to as "misclassification bias," refers to a distortion in estimation of the effect of interest that results when measurement of the exposure or the health outcome is systematically inaccurate.[78] The various sources of error that can lead to this bias include: the use of a measurement device that has a built-in or induced defect such as: a questionnaire, interview procedure, or index derived therefrom that does not measure what it purports to measure; an inaccurate diagnostic procedure for health outcome; or an incomplete or erroneous data source, e.g., as would occur when subject recall of exposure history is selective. Whatever the source of error in the information obtained, a study subject may as a result be misclassified in terms of exposure and/or health status. Selective recall of exposure history leads to a type of information bias known as "recall bias" and constitutes the most serious disadvantage of case-control studies. Stated differently, recall bias refers to the implication that exposure data may be influenced by experience of adverse outcomes, and thus must be suspected whenever interview data from affected and unaffected individuals are depended on as a primary source of information on exposure. Recall bias may produce either spuriously positive findings (cases overreport exposure, controls underreport exposure, or both) or spuriously negative findings (cases underreport exposure).[33] There are several ways to assess recall bias: by comparing self-reports of exposure with records of exposure collected prior to or independently of the outcome; by measuring exposure biochemically; or by comparing the frequency of self-reported exposure among several "worried" groups thought to have distinct conditions (such as chromosomally normal and chromosomally abnormal spontaneous abortions, neural tube defects, and Down's syndrome).[28]

Reporting bias. As noted by the 1981 March of Dimes Panel,[33] there are certain difficulties associated with investigating an aware and exposed population living in the vicinity of a waste site that extend beyond the methodologic issues that are statistically or epidemiologically tractable. For example, when community reports of adverse events are the impetus of subsequent health studies, statistical concerns arise that the hypothesis has been suggested by the data. Usual tests of significance may be inappropriate since one has chosen to test for an effect that has already been noted. While the observation, ideally, should be tested in another area, this may not be possible due to the uniqueness of the exposure or to other reasons that require the investigation to be confirmed. As advocated by the panel, when faced with this dilemma, approaches are needed which compensate for the positive bias without introducing negative bias. Excluding a reported cluster can negatively bias an exposure-outcome relationship if the community is small, the outcome is rare, and most of the current cases have been found. A prospective study on a truly independent sample could be mounted if the exposures were continuing; however, as discussed below, this design is generally not sensible for studying rare health outcomes. The March of Dimes Panel conceded that no truly satisfactory solution may exist to this dilemma, and investigators ". . . may be forced to include the original cluster in the analysis, relying on scrupulous definition of the denominator and unambiguous results to estimate the probability that the increase has occurred by chance."[33]

Interaction

Thus far, the general discussion of determining exposure-outcome relationships has devoted little attention to the problems of multiple concurrent exposures. As noted above, individuals living in the vicinity of hazardous waste sites may be exposed via many different routes to mixtures or combinations of several potentially toxic chemicals. When more than one exposure is involved, not only must the existence of distinct relationships between each exposure and health outcome be verified, but also the relationships among the exposures must be examined. For example, one exposure may confound another, leading to confounding bias in the estimate of the exposure-outcome relationship. In addition, the first exposure may interact with the second by potentiating or suppressing the relationship between the second exposure and health outcome. In this case the relationship of the second exposure to the outcome changes with the level of the first exposure.[32]

Several authors have considered the statistical/epidemiologic issue of interaction as it applies to the combined effect of two or more exposures.[81-84] Although the requisite analytic methods have been developed to assess interactions, their application to waste site studies may be limited since: the most useful and interpretable analysis of interaction requires the application of multivariate statistical techniques such as linear logistic regression,[75,78-80] and the typically weak and incomplete data derived from waste site studies may not be amenable to these more sophisticated modes of analysis; and several types of statistical models are available for assessing interaction but there is some dispute over which is the most appropriate.[82,83,85] The relevance of these statistical models to the biology of waste site-related illness will remain uncertain, however, until more is known about how various chemical exposures produce illness or biologic effects.[32]

SPECIFIC METHODOLOGIC APPROACHES TO EVALUATING HEALTH EFFECTS OF WASTE SITE EXPOSURES

The succeeding discussion of specific epidemiologic study designs is facilitated by first considering two general underlying themes developed by the March of Dimes Panel in their guidelines for reproductive effects studies:[33] the distinction between investigating a cluster of adverse effects (health outcome-based study) as opposed to investigating a supposedly hazardous exposure (exposure-based study), and the level of the investigation in terms of cost, complexity, and the time required for completion. While developed primarily for reproductive effects studies, these themes are equally applicable to the entire spectrum of health outcomes related to toxic waste site exposures.

Health Outcome-Based Versus Exposure-Based Studies

In cluster situations, the investigation is provoked by some real or apparent excess in the occurrence of an adverse health outcome. The first task is to determine the population at risk and to decide whether the observed spatial and/or temporal cluster of events represents a meaningful departure from the expected distribution in space and/or time. The second task, especially when there is no clue as to the cause, is to know which questions to ask in search of the cause.[33] Health outcome-based studies often take the form of a "crisis response" in which episodes of death or acute illness in small population groups spur attempts to establish the causative exposures so that appropriate treatment and control measures can be provided.[32]

In exposure situations, an episode involving known or suspected toxic substances has occurred and adverse effects are anticipated, although it may not be obvious what outcomes are salient for evaluating whether effects have occurred. Exposure-based studies are likely to be expensive and unrewarding unless specific toxic effects of the chemicals are known and exposures are well-defined and occurred at levels that could possibly lead to recognizable health effects.[33]

The Level of the Investigation

Based primarily on practical considerations, health effects investigations can be classified into three levels.[33] Level I is based on existing, routine, and easily accessible exposure and health outcome records. The investigation will usually be conducted with speed and economy and will seldom involve case examinations or special questionnaires. Level I studies will lack power since they will usually be limited to poorly defined measures of exposure. They may also be deficient by being unable to adjust estimates of exposure-outcome relationships for the effects of potentially confounding factors.

Since they generally involve aggregate versus individual data on exposure and health outcome, Level I investigations include the large class of ecologic studies. For example, Level I studies may draw on vital certificate data or special registries of tumors or malformations in order to examine birth weights, perinatal mortality, cancer incidence or mortality, or sex of offspring.

Level II includes short-term purposeful epidemiologic studies, such as cross-sectional, case-control, or short-term cohort, that require the collection of more precise, individual exposure and health outcome data as well as data on potentially confounding variables. The decision to proceed from the Level I descriptive evaluation to Level II analytic studies is based on the following criteria:[33]

1. A statistically significant association is observed at Level I but there is a need to explore a broader spectrum of outcomes and/or to specify more precisely the exposure-outcome relationship (e.g., threshold and dose-response components of the exposure).

2. The end point that stimulated the investigation, or that seems most biologically coherent with the exposure, is not accessible to Level I studies.

3. The results in Level I are inconclusive.

In Level II studies, the statistical considerations of power, bias, and interaction discussed above are applicable to the choice of study design, enabling the researcher to make maximal use of small numbers, rare events, and uncertain information sources. Level II studies can entertain a wide range of end points and can include outcomes identified in medical records (spontaneous abortions, malformations, behavioral or psychological disorders), through interviews with study subjects (spontaneous abortion, sexual dysfunction, symptoms or signs of rashes, paralysis, eye irritation), or through biological studies of study subjects (biochemical, immunologic, and chromosomal assessments, and nerve conduction velocities).[33]

Level III involves well-planned, long-term investigations such as prospective studies of exposed and unexposed residential cohorts. Since this design is well suited for diseases with long latency periods it has been considered mainly for the purpose of discovering environmental carcinogens.[33] Level III studies are greatly facilitated by the existence of centralized and accessible registries of births, deaths, and diseases such as the National Death Index[86] recently created by the National Center for Health Statistics.

An example of a current waste site-related investigation that encompasses all three levels of study is provided by CDC's proposed study of PCB-exposed cohorts.[31] This study proposes a systematic approach to evaluate the degree of human exposure and the extent of health effects at Superfund sites associated with elevated levels of PCBs. Their multistage approach, which is designed to detect significant human exposures with the least expenditure of cost and time, is divided into four stages:

1. Ecological assessment (Level I effort) to identify sites with PCB exposures

2. Pilot exposure study (Level II effort) to document body burdens of PCBs among the "most exposed" persons at each site

3. Community survey (Level II effort) to identify cohorts of PCB-exposed persons with little or no levels of other toxic chemicals which would confound the health effects of PCBs

4. Cohort study (Level III effort) to design and conduct registries of PCB-exposed cohorts detected in the third stage in order to examine the long-term health effects of low-level PCB exposure. Although Level III studies such as this are expensive, complex, and time consuming, they do provide one of the few tractable approaches to studying the long-term effects of low-dose exposures.

Classical Epidemiological Study Designs

This section highlights the advantages and disadvantages of several classical study designs that may be utilized to evaluate health effects at hazardous waste sites. Basic epidemiologic designs consist of three types of studies (given most emphasis in standard texts): cohort, cross-sectional, and case-control.[78,87,88] The objectives of the first two types may be descriptive or etiologic; whereas, the objectives of case-control studies are traditionally etiologic.[78] Also considered are certain "hybrid" or composite designs that combine or extend elements from the basic designs.[78] Both the basic and the hybrid approaches involve the individual as the unit of analysis. In addition, this section considers some potentially useful "incomplete" designs or studies in which information is missing on one or more relevant factors.[78] Included here are the broad class of ecologic models and a class of space/time chapter designs. Finally, the utility of population registries for long-term followup studies is discussed.

Basic Designs

Cohort Studies. A cohort study, also called a followup, incidence, panel, or prospective study, involves a design in which information about the study factor (exposed and unexposed) is known for all subjects at the beginning of the followup period. The population at risk of developing the health outcome, either disease incidence or mortality, is followed for a given period through reexaminations or population surveillance, during or after which new cases or deaths are identified.[78]

In cohort studies, the unexposed group may be defined as having no exposure to the agent under study (e.g., comparison of persons in an exposed community with persons in an unexposed community) or as exposed at lower doses than the exposed group (e.g., persons residing at varying distances from the site of environmental contamination). In the latter case, an effect may be missed if both low and higher doses of exposure share a common effect.[33] Ideally, in order to avoid bias, the exposed and unexposed cohorts should be identical with respect to: (1) the method of identification of the cohorts, (2) the method of followup, (3) the method of data collection on the health outcome and control variables, and (4) known risk factors for the health outcomes under study.[33] It is possible in cohort studies to enhance statistical power for a given significance level and effect size by increasing the unexposed group size relative to the exposed group size.[74]

A cohort study may be conducted completely prospectively, retrospectively, or ambispectively (combination of retrospective and prospective) with respect to health outcome. Since it most closely resembles an experimental situation, the prospective design is generally preferred for making causal inferences. The retrospective cohort study, however, is usually more feasible for studying rare diseases or diseases associated with long latent periods. The applicability of the retrospective design, of course, depends on the availability of previous study factor information on a well-defined population that has been followed for detection of new cases or deaths.[78] Since the requirements of a retrospective cohort study can often be met in industrial settings, this design has served and continues to serve as a major analytic framework for studies of occupational mortality and morbidity.[89-92]

The issue of followup of an identified population is a fundamental concept that impacts on the methodology of any retrospective cohort study. Followup methods associated with population-based studies have been well documented in the literature.[93-97] A particularly relevant reference regarding waste site studies is the report by Austin et al.[97] on the long-term followup of a specific community-based or residential cohort.

The advantages of the general cohort design over other designs for evaluating health effects at hazardous waste sites include: (1) the exposure level on each subject is observed at the onset of the followup period before the health outcome is detected. Thus, it is reasonably certain that the hypothesized cause preceded the health outcome, and that outcome status did not differentially influence the selection of subjects by exposure level; (2) for a given effect size, level of significance, and power, the cohort design will require a smaller sample size than the case-control study when the exposure is rarer than the outcome; (3) the relative risk is directly estimable as a measure of association; (4) compared to case-control designs it is possible to obtain a more precise measure of exposure; and (5) unlike case-control designs, a wide variety of health outcomes can be examined within a single study.[33,78]

The chief weaknesses of the cohort study design are: (1) loss of study subjects because of migration, lack of participation, or death (or competing causes of death in mortality studies). This is a potential problem in any type of cohort study, particularly those involving disease incidence and long-term followup. Such systematic study population attrition can lead to certain biases that cannot be controlled during analysis; (2) the cohort design is statistically and practically inefficient for studying rare diseases; (3) the retrospective cohort (or historical prospective) study, in particular, may not be feasible for health evaluations of waste site exposure due to the inability to precisely define or locate demographic data for an historical "residential" cohort of individuals who lived in the vicinity of the waste site, the inability to classify the type and degree of historical exposures, the difficulties related to the size of the population, the selection of an appropriate historical comparison population, and ascertainment of health outcome status between the point of exposure and the end of the followup period.[33,78]

Because the relatively stringent requirements of the cohort (or retrospective cohort) design preclude its ready application to many waste site investigations, it is not surprising that this design was not used in any of the published waste site studies that are reviewed in this chapter. There are, however, currently several attempts planned or underway to utilize this design in waste site-exposed environments where the population at risk can be identified and followed, and surveillance of specific diseases can be done. The most promising of these approaches is the previously mentioned prospective study of PCB-exposed cohorts planned by CDC.[31] This study has a good chance to succeed since it is strictly prospective in nature, the populations at risk are well defined, and exposures are well documented.

Another promising approach, involving the historical-prospective design, is the proposed mortality study of Talbott and Radford[98] of a community exposed since 1910 to low-level radon and gamma radiation. This study will attempt to evaluate the total and cause-specific mortality experience of a community cohort previously identified by Talbott et al.[99] via the cross-sectional approach as having a higher rate of radiation-related thyroid diseases as compared to residents of a nearby unexposed community. Their proposed mortality study includes an exposed cohort of 6000 persons living since 1938 within a one-mile radius of a uranium waste site (near Canonsburg, Pennsylvania) and a control cohort of 6000 persons living at the same time in a nearby unexposed community (Bridgeville, Pennsylvania). Study subjects will be identified from capitation and property tax rosters (containing information on name, address, and occupation) that are maintained by the local townships. Using vital status information available from next-of-kin, neighbors, driver license bureaus, and other sources, the two residential cohorts will be followed for deaths through a recent date such as 1983. The success of this study will depend largely on the ability to trace the cohorts in the absence of social security numbers (an important record linkage key), which are not available from the tax rosters.

In addition to the aforementioned studies, other historical-prospective or strictly prospective designs are being considered by others such as state health departments to evaluate health effects of waste site exposures. For example, the Pennsylvania Department of Health has developed a protocol to systematically investigate the health status of former employees and selected residents of Lock Haven, Pennsylvania, who may have experienced hazardous exposures associated with the Drake Chemical Superfund site.[100] This protocol includes both a conventional retrospective occupational cohort as well as a current and retrospective community-based cohort of households in the immediate vicinity of the Drake site. For comparative purposes, a "control" cohort consisting of a similar number of households will be randomly selected from the Lock Haven area. In addition to a mortality/morbidity followup of the occupational and residential cohorts, the Drake site protocol includes a cancer mortality and congenital malformation incidence review, a health questionnaire survey, and a bladder cancer screening component. Further details of this protocol are provided in a subsequent section.

As described below there are also a number of retrospective or prospective cohort designs in other states being considered as approaches to evaluating health effects at waste sites.

Cross-Sectional Studies. A cross-sectional study, also called a survey or prevalence study, involves a nondirectional or backward design of a study population that has been selected from a single target population. This type of design involves the prevalence of health outcomes rather than the incidence, and usually involves random sampling of the target population. The backward design begins with the classification of disease or dysfunction (e.g., case vs noncase) and proceeds by obtaining through interview or examination information about individual histories of the study factor, (i.e., previous exposures, events, or characteristics). In contrast, with the nondirectional design both the study factor and the disease are observed simultaneously, so that neither variable may be uniquely identified as occurring first. The utility of cross-sectional studies for describing the frequencies of health outcomes or other characteristics and for making causal inferences is severely limited if random probability sampling is not incorporated into the design.[78]

Since the cross-sectional design does not involve a followup period, it is often used to generate new etiologic hypotheses regarding study factors and/or health outcomes. Cross-sectional studies are particularly useful for studying conditions that are quantitatively measured and that can vary over time (e.g., blood pressure) or relatively frequent diseases that have long duration (e.g., chronic bronchitis). They are not appropriate for studying rare diseases or diseases with short duration. Because the results of cross-sectional studies are usually largely derived from interview data, they are especially prone to the methodologic limitations associated with nonresponse, nonspecificity of health outcome and/or exposure, recall bias, and the necessity of working in a highly charged emotional atmosphere.[78]

Despite its many limitations, some form of the cross-sectional design incorporating questionnaires has been utilized in 10 of the 14 waste site-related health studies that were reviewed in this chapter.

Case-Control Studies. The case-control study, also called a case-referent, case history, or retrospective study, involves a backward or nondirectional design that compares a group of cases and one or more groups of noncases (controls) with respect to a current or previous study factor level (exposure).[78] A fundamental difference between this study design and the cross-sectional design is that the study groups in the case-control design are selected from separate populations of available cases and noncases rather than from a single target population. However, to make causal inferences it must be assumed that the control subjects are representative of the same candidate population from which the cases were developed. The control group may be derived from a number of sources including hospitals, neighborhoods, or the general population from which the cases were identified. Moreover, in order to ensure the validity of exposure assessments (e.g., via interview) and concurrently to avoid missing an association where exposure is related to more than one health outcome, an investigator may choose to select two or more control groups; one defined by an adverse health outcome other than that being studied and the other unaffected by any adverse outcome. Another difference is that case-control studies, unlike cross-sectional, may involve cases and noncases identified over time (i.e., incident cases) as in a cohort study.[33,75,78,79]

Ideally, cases and controls should be comparable with respect to: (1) the a priori probability of exposure, (2) the method of ascertainment, (3) the method of collection and the reliability and validity of data on exposure status and potentially confounding variables, and (4) all characteristics (other than the study factors) that relate both to the health outcome under study and to the exposure variable (i.e., confounding variables).[33] Since it is usually not possible at the outset of a study to ascertain the comparability of cases and controls with respect to potentially confounding variables, efforts are generally made to control for confounding bias either through design (matching cases to one or more controls on the basis of one or more confounding characteristics) or through analysis (stratification by levels of one or more confounding characteristics). The power of a case-control design can be increased (for a given significance level and effect size) by choosing two or more controls for each case.[33,78]

There are numerous practical and statistical advantages to case-control studies over other designs that include: (1) they are well-suited to testing etiologic hypotheses for specific rare diseases, (2) they allow for the investigation of diseases with any latent period or duration of expression, (3) since the ratio of cases to controls can be fixed by the investigator, analyses are more statistically efficient than they are for other designs for a given sample size and study cost, (4) the convenient sampling strategy and the relatively short study period usually make case-control studies less time consuming and expensive than cohort or cross-sectional studies, (5) for a given effect size, level of significance, and power, the case-control design will require fewer individuals than the cohort study when the studied health outcome is rarer than the exposure, and (6) compared to a cohort study, the case-control study may permit a more precise clinical classification of the cases.[33,78]

The principal limitations of case-control studies compared to other designs include: (1) the ability to distinguish antecedent events from consequent events (i.e., to make causal inferences) depends on the retrospective ascertainment of exposure information from records, which may be inaccurate or incomplete, or from human recall, which is subject to differential error for cases and controls (i.e., recall bias), (2) only one health outcome of interest can be entertained within a particular case-control study and this outcome must be measured as a categorical variable. Thus, the case-control design may not be appropriate for exploring the range of health effects resulting from a certain exposure, (3) without additional information, this design is not appropriate for estimating the prevalence of a disease in a population, and (4) the measure of association in case-control studies is the odds ratio which only approximates (although quite well for rare outcomes) the relative risk that derives from cohort studies.[33,78]

To date the classical case-control approach has not been widely applied to evaluate health effects of waste site or other toxic environmental exposures, however, there have been applications of a hybrid case-control/cohort design as discussed below.

Hybrid Designs

Of the several hybrid designs described by Kleinbaum et al.,[78] there are five that may be particularly useful for evaluating health effects of toxic waste site exposures.

Ambidirectional. The ambidirectional study, known also as a nested case-control study combines a few of the major advantages of both cohort and case-control studies. In this design, a single population is defined at the onset without regard to exposure information and is followed for a given period for the detection of all incident cases or deaths. The incident cases or deaths are then compared with a group of controls sampled from the same population with respect to previous or current exposure levels. The controls may be sampled randomly from the population, or they may be matched to the incidence cases or deaths. The nested case-control design is usually applied when an etiologic hypothesis emerges after the beginning of followup or when limited resources preclude the measurement of exposures on every subject in the study population.

The major advantage of the ambidirectional design over case-control designs is the assurance that cases and controls are identified from the same well-defined population. Furthermore, since exposure information in ambi-directional studies is obtained only on a small fraction of the noncases in the study population, this design, unlike the cohort design, is suitable for studying rare diseases.[78]

An appropriate situation for an ambidirectional study is one in which it is possible to identify most new cases (or deaths) of one or more rare diseases in a large population by using existing information systems such as employment or insurance records, a disease registry, or vital records.[78] An example of such an application is the work of Lyon et al.[101] who studied cancer clustering around a coke oven and uranium tailing dump. In this study, the distribution of distances to the point source of exposure (i.e., the exposure variable) for cases of lung cancer in a two county area between 1966 and 1975 was compared to the distribution for a control group of other cancer cases that occurred in the same area and time period. Both the cases and controls were drawn from the Utah Cancer Registry.

Although not related to hazardous waste site effects, a similar approach was also recently taken by Blot et al.[102] who compared the occupational histories of lung cancer cases and controls identified from Pennsylvania vital statistics records.

Prevalence Study. A followup prevalence study combines the elements of the cohort and cross-sectional designs by following a population over time whose exposure status is known at the onset but whose health outcome status is not known until a subsequent examination is conducted. This design, which involves only prevalent cases, is most useful when, at the planning stage, the health outcome eventually measured is not considered, or is not known, as a possible consequence of the exposure.[78]

Selective Prevalence Study. A selective prevalence study is similar to the followup prevalence study except that only a subset of individuals who are potentially eligible for developing the health outcome of interest are identified from the original population and followed over time. This type of design might facilitate the etiologic investigations of certain waste site exposure-related health outcomes (e.g. birth defects) for which the candidate population cannot be readily identified.[78]

The remaining hybrid designs described below that might be applicable to waste site situations essentially extend the strategy of one basic design through repetition over time.

Repeated Survey. In a repeated survey, a sequence of two or more independent cross-sectional studies are conducted within the same dynamic population. It is unlikely, therefore, that this design will include repeated observations of the same study subjects. This design can be applied to assess overall health status changes in large populations and to determine how observed trends in health outcome relate to changes in exposure levels.[78]

Repeated Followup. A repeated followup study, or a repeated measures study, involves a fixed cohort study in which the total followup period is divided into two or more subperiods. This design is generally implemented by conducting three or more examinations and/or interviews spaced months or perhaps years apart, resulting in the collection of exposure and health outcome status information (repeated measures) on all participating subjects at each exam. This design is particularly useful for investigating the possible association between exposure change and health outcome events, for investigating the etiology and natural history of remittent diseases, and for investigating the relationship between two health outcomes.[78]

Incomplete Designs

Incomplete designs, being Level I investigations, are frequently used when data are not readily available for conducting another type of study. As evident below, it is often relatively inexpensive or convenient to utilize secondary data sources to test or generate etiologic hypotheses using these designs before considerable time and resources are allocated to primary data collection. This section considers three of several classes of incomplete designs described by Kleinbaum et al.[78] for potential application to hazardous waste site health effects evaluations. In addition, reference is made to several secondary existing data sources that may be incorporated into these designs.

Ecologic Studies. Broadly defined, ecologic studies, also called aggregate or descriptive studies, are empirical investigations involving the group as the unit of analysis. Typically, the group is a geographically defined area such as a state, county, or census tract. Ecologic analysis may involve incidence, prevalence, or mortality data, but the latter is most common because of the widespread availability of such data.[78] The primary analytic feature of an ecologic study is the lack of information about the joint distribution of the study factor and the disease within each group (i.e., unit of analysis).

In his recent review article, Morgenstern[103] distinguishes among four basic types of ecologic studies:

1. The *exploratory study* is the simplest type in which geographic differences in the disease rate are observed among several regions (groups). The objective is to search for spatial patterns that might suggest an environmental etiology or a special etiologic hypothesis. No exposures are measured, and, generally, no formal data analysis is used. For example, using detailed mortality data assembled by NCHS, the National Cancer Institute mapped the age-adjusted cancer mortality rates in the Unites States by county and state for the period 1950 to 1969.[104]

2. In the *multiple-group comparison study*, the association between the average exposure level and the disease rate is observed among several groups at one point in time or during one period. Using statistical regression techniques, it is possible with this design to estimate the same parameters that are estimable at the individual level for quantifying the magnitude of association between the exposure and the disease.[105-106] However, as discussed below, the ecologic estimates of categorical measures are subject to biases unique to grouped data.

3. In the *time trend study*, the relationship is observed between the change in the average exposure level and the change in the disease rate for a single population. For time trend studies involving a sudden change in exposure, the slope in the disease trend is compared before and after the change, whereas, for those involving a gradual change in exposure, the time trends for both exposure and disease must be compared.

4. In the *mixed study*, design types (2) and (3) above are essentially combined; i.e., the relationship is observed between the changes in average exposure level and disease rate among several groups. In general, compared to the other designs, the mixed design provides a better test of an etiologic hypothesis since the results of the mixed study are less likely to be due to the confounding effects of extraneous risk factors.[103]

Since ecologic studies can often be done by combining existing data files on large populations, they are generally less expensive and less time consuming than studies involving the individual as the unit of an analysis. Ecologic studies are, therefore, well-suited as a preliminary or exploratory approach to evaluating health effects of waste site exposures. However, despite these practical advantages, causal inference about individual events from grouped data is limited by the following methodological problems:

1. Data on many extraneous risk factors (age, sex, occupation, personal habits) may not be available on the ecologic level, thus preventing control over confounding bias on the observed exposure-health outcome relationships. For example, geographic variations in mortality are probably more likely due to differences in socioeconomic status, alcohol and tobacco consumption, and diet rather than to any common source waste site exposure unless that exposure is very intense or exceptionally toxic.

2. The populations at risk very often leave the study (i.e., community exposed to toxic waste site materials) prior to the investigation so that examining cross-sectional data with ecologic analysis (e.g., a multiple group comparison study) is often fallacious.[72]

3. A particular limitation with ecologic time trend analysis is that generally only mortality data are available. Morbidity data, when available, are difficult to interpret due to either changes in ascertainment or changes in definitions of disease. Moreover, trend data can be greatly affected by extraneous factors such as the prevalence of chronic disease in the population (an indicator of susceptibility to effects of toxic exposures) and also more powerful factors such as weather, respiratory infections, and natural disasters.[72]

4. In ecologic analysis, there is the potential for substantial bias in effect modification. This problem, known as the ecological fallacy, occurs when the composition of each group is not homogeneous with respect to the study factor.[107] Theoretically, the bias resulting from ecologic analysis can make an association appear stronger or weaker than it is at the individual level; however, in practice, this bias ordinarily exaggerates the magnitude of a true association, if one exists.[108-110]

5. With ecologic analysis, certain predictor variables (especially sociodemographic and environmental variables) tend to be more highly correlated with each other than they are at the individual level — a phenomenon called "multicollinearity."[103] Consequently, the increased correlations between predictor variables make it particularly difficult to isolate their effects on the health outcome. In general, multicollinearity is most problematic for ecologic studies Involving larger and fewer geographically defined units of analysis.[109-110]

These and other problems with ecologic analysis such as measurement error and ambiguity of cause and effect can be minimized, however, via the following approaches:[103] (1) use of ecologic regression rather than correlation including in the model as many risk factors as possible, (2) use of data that are grouped into the smallest geographic units of analyses as possible, subject to the constraints of intergroup migration and unstable rate estimation, and (3) attempt to ascertain how the groups were formed and analyze using all variables thought to be related to the grouping process.

Available Secondary Databases For Ecologic Studies

In addition to the strengths and weaknesses described above, the utility of ecologic analysis for evaluating health effects at hazardous waste sites depends heavily upon the availability of published summary data on exposure and/or health outcome that are specific for an appropriate unit of analysis. The National Priority List databases and the centralized toxicological databanks for example, are national level data repositories that may provide useful summary data on potential exposures specific to geographic areas that contain toxic waste sites.

On the other hand, the availability and accessibility of ecologic (or individual) data on health outcomes relevant to waste site-exposed areas varies according to geographic area and type of outcome. No attempt will be made in this chapter to provide an exhaustive listing of data sources for all of the health outcomes discussed above. This section does, however, describe some of the larger sources of data on morbidity and mortality that may be useful in the design and conduct of ecologic analyses.

The National Center for Health Statistics (NCHS) publishes summary vital statistics data collected through states on numerous topics. This is a particularly good source for determination of state, metropolitan, and national birth and death rates. Moreover, much of the NCHS data are available at the detailed individual record level on magnetic tapes which can be purchased through the National Technical Information Service.[111]

To overcome the lack of specificity inherent in published mortality rates (e.g., age-specific death rates at the county level are not published) several institutions including the University of Pittsburgh Graduate School of Public Health and the Johns Hopkins School of Hygiene and Public Health have linked the NCHS detailed mortality data with detailed U.S. census population data to develop computerized data retrieval/rate generating systems.[112-113] For example, the Mortality and Population Database System (MPDS) developed by Marsh et al.[112] at the University of Pittsburgh can generate death rates according to state, county, age, race, and sex for the years 1950 to 1979 for any cause of death (cancer deaths only for 1950 to 1962) specified by the appropriate four-digit International Classification of Diseases (ICD) code.

The availability of ecologic data on morbidity, in particular cancer incidence data, is much more dependent upon the geographic area of study. For example, as mentioned above cancer incidence data developed through NCI's Surveillance, Epidemiology, and End Results (SEER) Program is available for certain years for only about 10% of the U.S. population who reside in the major Standard Metropolitan Statistical Areas (SMSA).[114] Cancer incidence data for other geographic areas such as states, counties, or localities are available only for those states or subdivisions thereof that have developed tumor registry programs. Currently, tumor registries exist or are under development in about twelve states. Greenberg et al.[115] provides an extensive review of the measurement, sources, and uses of cancer incidence data in the United States.

Data systems are also available that integrate mortality and morbidity statistics with various sources of environmental data. The data included on two such systems, the Socio-Economic-Environmental Demographic Information System (SEEDIS)[116-117] and UPGRADE[118-119] are described in a review article by McCrea-Curnen and Schoenfeld.[120] This article provides an excellent review of national cancer mortality and morbidity data sources as well as a description of other systems that are presently available for linking cancer morbidity and mortality with information on environmental factors.

The remaining approaches to evaluating health effects described in this section involve the individual as opposed to a group as the primary unit of analysis.

Proportional Studies. A proportional (or proportionate study) is one that only includes observations on incident cases or deaths without information about the candidate population at risk of developing the health outcome(s).[78] Due to the availability of mortality data, the proportional mortality design has been more widely applied, particularly in studies of occupational groups.[121-122] Proportional studies may be viewed as a special type of cross-sectional study[123-124] or as a special type of case-control study.[125] In either case, the basic approach of the proportional study is to compare the proportion of total cases (or deaths) resulting from the disease of interest among different levels of exposure. From this approach, therefore, it is only possible to test the exposure-outcome relationship of primary interest if it can be assumed that there is no association between the exposure variable and the remaining (or comparison) diseases. Due to the limitations associated with using mortality and most morbidity (i.e. overt disease) data in health effects evaluations of waste site-exposed communities, the utility of the proportional design for such studies is, therefore, restricted.

Space/Time Cluster Studies. In this class of incomplete designs, two or more supposedly nonrandom aggregations or clusters of disease (incidence, prevalence, or mortality) are observed, differentially distributed in space, over time, or together in space and time.[87] The cases may be linked to a well-defined population; but no exposure data need to be measured or collected. *Space clustering* is a nonuniform distribution of cases over the total study area, relative to the distribution of the candidate population and is used by epidemiologists to suggest an environmental etiology of the health outcomes.[78] The simplest type of space clustering is the aforementioned ecologic exploratory study. The cancer clustering study of Lyon et al.,[101] although not an incomplete design, can be considered as a type of space clustering design.

Time clustering is a nonuniform distribution of cases over the duration of the study for a given candidate population.[78] Incomplete trend studies are essentially crude ecologic analyses that lack exposure information. In the context of waste site studies, this design may be used to identify and investigate local epidemics of adverse health outcome(s) or to identify cyclic fluctuations in their occurrence. Statistical tests are available for determining whether an observed cyclic variation is simply a chance occurrence.[126]

Space/time clustering, which involves the interaction between place and time of health outcome occurrence, is an approach that can be used to strengthen an exposure-outcome inference, particularly of an infectious etiology. Statistical procedures are also available for assessing the significance of space/time clusters.[127]

Establishment of Registries of Potentially Affected Persons

The establishment of registries of persons possibly exposed to toxic waste site materials is, in principle, similar to the determination of the exposed and unexposed groups in a cohort study. As with the cohort design, a registry may be established currently with prospective followup or established retrospectively with current or prospective followup. There are, however, two basic distinctions behind the two approaches.

First, in the cohort study the exposure status of each person is known at the onset of the study. Exposure status of persons enrolled in a registry may or may not be known until subsequent examinations and/or interviews are conducted. Registries developed without documented exposure information may be inefficient, however, for they may turn out to include a higher unexposed to exposed ratio than is needed for an acceptable statistical power to detect an excess in adverse health outcomes.

Second, exposure and health outcome data collected on subjects in a cohort study may not be maintained or updated after completion of the study. That is, incident cases or deaths are usually identified for the specific study at hand that is conducted within a given predetermined time period. The registry on the other hand, can be considered as an open-ended cohort study. That is, a registry provides a general database of exposed and unexposed persons that can be exploited in numerous ways to determine possible consequences of chemical or other exposures. For example, the registry data are likely to be amenable to cohort cross-sectional or case-control designs and perhaps most importantly they facilitate application of the more useful hybrid designs such as the repeated survey or repeated followup. Furthermore, with a registry it is theoretically possible to maximize the available information on exposure-health outcome relationships by extending followup until all participants have had adequate time to succumb to the outcomes of interest.

In general, therefore, with a registry it is not necessary to develop a specific study protocol until after the potentially exposed persons have been identified. This characteristic is most advantageous since with the rapid mobility of the general population, the identification of persons possibly exposed to the hazard present at a waste site must be made as soon as possible following recognition of the problem. Thus, ideally, if a particular waste site is suspected of posing a hazard to human health, a registry of potentially affected people should be established as one of the first courses of action.

The preceding discussion has focused primarily on the notion of a site-specific registry, i.e., a registry established with persons possibly exposed to material from a particular waste site. Such a registry will likely involve one localized, relatively homogeneous population that has potential exposure to a wide variety of waste site materials. In contrast, an exposure-specific registry is one which assembles persons from two or more locations on the basis of their common exposure to one hazardous material. While such registries are homogeneous with respect to exposure, they are more likely to be much less homogeneous with respect to other factors that might potentially confound an exposure-outcome relationship. An example of an exposure-based registry is the registry of PCB-exposed persons recently proposed by CDC.[31]

The establishment of a registry can also provide crucial information needed by various state and federal agencies who may recognize the need for epidemiologic studies for research or for determining who might require health care and long-term followup.[2] It is also important to recognize that not all waste site situations are amenable to or even require the establishment of registries. In particular, the long-term usefulness of registries may be restricted by the inability to locate persons who have migrated out of the study area. Although there are centralized federal, state, and local sources that can be utilized for tracing individuals (e.g., Social Security Administration, state driver license bureaus), they may require key record linkage elements (e.g. social security numbers) that may not be available from existing record sources. As previously mentioned, certain of the newly developed automated data systems such as the National Death Index may ultimately determine the success of existing or planned registries. As shown below, the federal government and several state health departments have initiated the establishment of registries of persons possibly exposed to hazardous waste site materials.

Alternative or Nonclassical Approaches

It has been shown throughout this report that the application of classical epidemiological methods to evaluate health effects at hazardous waste sites is made difficult due to a wide variety of methodological limitations and particularly complex real-life situations. In view of this dilemma, it is crucial that environmental epidemiologists begin both to develop methods to enhance the analytic capabilities of the classical approaches and to consider alternative methodologic approaches that will pave the way to a more complete understanding and perhaps an ultimate solution of waste site-related health problems. In general, there are four categories of "nonclassical" approaches that might be pursued:

1. To increase the inferential capabilities of existing statistical and epidemiologic methods by increasing analytic control over extraneous factors (e.g., multivariate methods) or by decreasing the dependency of the methods to underlying assumptions or requirements (e.g., development of nonparametric alternatives)

2. To explore familiar roles for epidemiology in nonenvironmental settings and to utilize these roles as paradigms for possible roles in hazardous waste site settings

3. To consider other nonepidemiologic methods that are used to assess analogous problems in nonenvironmental settings

4. To employ the classical methods of epidemiology to study the health experience of occupational groups that are heavily exposed to substances commonly found in waste sites

Although the statistical and epidemiologic literature abounds with activities and developments related to the first category of enhanced classical approaches, there is inevitably a prolonged lag before such refinements become a routine component in actual research problems. For example, in recent years there has been considerable growth in the body of knowledge related to case-control methodology (e.g., linear logistic regression techniques, log linear modeling, and proportional hazards modeling); however, most of these newer methods require a level of mathematical and computer programming sophistication that impedes their rapid dissemination and application to real-life problems. Researchers in environmental epidemiology should make concerted efforts to regularly review the relevant literature in order to expeditiously exploit, to the fullest extent possible, any new methodologies that could be brought to bear on hazardous waste site epidemiology.

The second general category of alternative approaches was discussed by Neutra at the 1981 Rockefeller Symposium.[128] Arguing by analogy of the function of epidemiology in the infectious disease field, Neutra relates a paradigmatic model for the natural history of infectious disease to an analogous model for the natural history of chemically induced illness.

At the same symposium, Selikoff also endorsed the need for this second category of alternatives by advocating the development and application of approaches such as seroepidemiology, biochemical epidemiology, and epidemiological immunotoxicology to health effects evaluations at waste sites.[129]

Compared to the first two categories, the third category of alternative approaches (i.e., nonepidemiologic methods) to waste site-related health evaluations has probably received the least attention in the scientific literature. Neutra[128] also alludes to this category by suggesting, for example, that data on subjective symptomatology (which is prevalent in health surveys of waste site-exposed communities and is often viewed as psychosomatic or hypochondriacal) be subjected to nonclassical epidemiologic techniques such as numerical taxonomy[130] (a type of cluster analysis) and discrimination function analysis or principle components analysis[131] in order to assess whether patterns of simultaneous symptoms differ for exposed and unexposed groups. In an analogous fashion, the techniques of numerical taxonomy and discriminate analysis have been used quite successfully in the epidemiology of colitis[132] and other poorly understood syndromes to assess whether reported complaints constitute any recognizable syndromes.

The fourth category of alternative approaches is fundamentally different from the others since it does not directly involve the waste site-exposed community but rather a surrogate target population that has received similar, albeit, more intense exposures. Occupational groups, in general, are especially amenable to epidemiologic inquiry since historical records are often available that permit the construction of well-defined cohorts which can be studied for diseases that are relatively rare and/or are associated with long latency periods. Moreover, since historical exposures received by cohort members are often documented or can be adequately extrapolated from current measurements, the health outcomes among occupational cohorts can often be related to type, duration, and intensity of exposure(s).

Thus, the study of an occupational group that has received relatively heavy exposure to a waste site-related agent of interest enables the determination of possible biological end points and the examination of dose-response relationships so that at least an effort at extrapolation to the community population can be made.

The utility of this fourth approach has been recognized by many investigators. For example, the Pennsylvania Health Department epidemiologists who have included both an occupational and community cohort in their proposed study of the Lock Haven, Pennsylvania Superfund site.[100] Perhaps the ideal occupational groups for study, however, are those whose work involves direct exposure to waste site materials, such as equipment operators who clean or maintain waste sites or workers aboard incinerator ships at sea.[27]

The continued development of all categories of effective nonclassical approaches will require, at a minimum, increased communication and collaboration among researchers from a variety of allied professions, including epidemiology, medicine, biostatistics, mathematics, engineering, and toxicology. The three recent published meetings,[36-38] as well as other chapters in this book have provided the groundwork for future collaborative efforts in the hazardous waste site field.

WASTE SITE-RELATED RESPONSIBILITIES AND RESEARCH ACTIVITIES AT THE FEDERAL, STATE, AND PRIVATE LEVELS

This section discusses the responsibilities and activities of federal and state governments and of independent researchers regarding adverse health effects caused by hazardous waste sites. Example studies were sought to exemplify these activities. An effort was made to gather published epidemiologic studies by systematically reviewing *Index Medicus*, an index of medical literature, from 1972 to the present. Two computer database indices, *Enviroline* and *Toxline*, were also searched for relevant references in the years 1980 to 1983. Examples of studies done by state health departments were obtained by contacting various state environmental epidemiologists. Protocols for ongoing or planned studies were also included if they serve to further illustrate the current epidemiologic methods being used to investigate health problems around hazardous waste sites.

Many of the studies reviewed and summarized in this section have been cited elsewhere in this chapter to illustrate various methodologic or other aspects of hazardous waste site health evaluations.

Federal Government Activities

Federal agencies regulate the handling, transportation, and storage of hazardous wastes. They have responsibility for overseeing state regulation of hazardous wastes and they conduct their own research. This section discusses activities of the Environmental Protection Agency (EPA) and the Centers for Disease Control (CDC) as well as mentioning other agencies.

Environmental Protection Agency

The EPA's role is basically environmental control, while CDC is more directly involved with the health of people living around hazardous waste sites. As briefly described in the introduction to this chapter, the authority to regulate hazardous wastes comes from several legislative acts. The Resource Conservation and Recovery Act of 1976 (RCRA) deals with the tracking and regulation of hazardous wastes. The Comprehensive Environmental Response, Compensation and Liability Act of 1980 (CERCLA or "Superfund") provides a comprehensive and effective way to deal with abandoned waste sites—both emergencies and long-term cleanup efforts. The EPA can also use the authority under other acts such as the Clean Water Act, the Safe Drinking Water Act, the Clean Air Act, the Toxic Substances Control Act, and the Refuse Act.

RCRA gives EPA the authority to develop a nationwide program to regulate hazardous wastes. EPA has developed regulations for:[133]

- Identification and listing of hazardous wastes

- Standards for generators of hazardous wastes

- Standards for transporters of hazardous wastes

- Standards for owners and operators of hazardous wastes treatment, storage, and disposal facilities

- Requirements for the issuance of permits to hazardous wastes facilities

- Guidelines for authorizing state hazardous wastes programs

Under RCRA the EPA can initiate legal action to require responsible parties to clean up a site that presents an "imminent and substantial" danger to health or the environment. Under CERCLA, the EPA can undertake cleanup when the responsible party or local governments cannot or will not do so. The CERCLA plan requires EPA to interact with local communities on any action lasting more than two weeks; the EPA Regional Office and the involved state must develop community relations plans to inform the public and work with local authorities.

The EPA provides guidance and information to states through many programs. A training program for local authorities has been developed by EPA to train personnel in areas such as hazardous environmental response; hazard evaluation, mitigation, and treatment; and personnel protection and safety.[133] A computer system, the Chemical Substances Information Network (CSIN; see Appendix A), links many computer data systems together providing information on chemicals' toxicity, production, health and environmental effects, and other aspects.[26] It was developed jointly by the EPA and the Council on Environmental Quality (CEQ) in accordance with the Toxic Substances Control Act. Another computer database, the Computer Aided Environmental Legislative Data System (CELDS), describes federal and state environmental legislation (rules, laws, regulations, standards).

An example of planned research by the EPA is the dioxin program designed to determine the extent of nationwide contamination, assess human health risks, and develop cleanup methods.[134] It involves a massive sampling of sites across the country, testing for dioxin contamination. The EPA will decide whether people should be evacuated if sampling shows an excess concentration. The EPA will continue to work with other agencies like CDC to research long-term health effects, which are unknown.

The EPA in conjunction with the Mitre Corporation of McLean, Virginia has also developed a computerized databank of selected technical data from the National Priority List (NPL) and other sites. This databank may be useful as a basis for planning future health evaluations. As discussed above, a National Priority List of 539 hazardous waste sites was proposed in August 1983 as the result of the efforts of states, EPA regions and headquarters to identify those sites of highest priority for remedial action. Selection of sites was based primarily on scores derived from the EPA Hazard Ranking System (HRS) developed in 1982 by the Mitre Corporation. The HRS scores, the supporting documentation packages for the scores, and the related EPA site inspection reports (Form 2070) are routinely submitted by the ten EPA regions as part of a public docket of technical data to support the decision to list these sites.

Using the Statistical Analysis System (SAS),[135] the Mitre Corporation has coded and entered selected technical data from this docket into an automated data management system known as the NPL Technical Data Base (TDB). As of early 1984, the database contained information for about 1000 sites including the 539 NPL sites.[136]

Technical data contained on the NPL-TDB include the site identification numbers, the site names and locations; the fire, explosion, and direct contact route scores,[137] the HRS scores for all sites; the groundwater, surface water, air route scores; and the individual rating scores under each of the three routes.

In addition, the NPL-TDB includes the raw data (e.g., the actual distance to the nearest well and the number of drums on site) on which the HRS scores were based.

Regarding exposure data, the NPL-TDB contains a list of up to 15 chemicals found at each site with flags indicating which chemicals were observed in groundwater, surface water, or air at the site, which were used to score toxicity in each of the three pathways for the HRS, and whether or not EPA's docket file contains concentration data for that chemical.

Appendix B provides a partial list of specific data elements stored in the NPL-TDB. At the present time, access to the NPL database is limited to Mitre staff responding to requests from EPA's Office of Emergency and Remedial Response.[136]

Centers for Disease Control (CDC)

CDC has five basic areas of involvement:

1. *Health risk assessment involving abandoned waste sites.* The National Institute for Occupational Safety and Health (NIOSH) becomes involved when the health of workers at the site is in question.

2. *Emergency response.* A toxic waste spill is one example.

3. *Specific health studies.* These are studied with outside funding and include new tests for chemicals that measure health risks, new health indicators, and the dioxin study mentioned above.

4. *Worker safety and health.* NIOSH has the lead in this area.

5. *Data collection*. For example, building databases on specific chemicals is one activity. The National Library of Medicine and the National Toxicology Program have the lead in this area.

CERCLA guides activities to remedy problems around abandoned waste sites. Under provisions of the Act, the first step is for EPA and states to identify existing sites. Sites are then ranked by types and quantities of toxic chemicals and by potential threat to human health. CDC assesses the potential for adverse health effects and makes recommendations for the protection of public health. If CDC decides a study is appropriate, it might design the study and submit a request to EPA for funding. State health departments are involved in the beginning of any study and sometimes initiate them. Like the EPA, CDC provides information for state health departments. CDC published a document called *"A System for Prevention, Assessment, and Control of Exposures and Health Effects from Hazardous Sites."*[138] This gives step-by-step procedures that aid in decisionmaking and planning for investigations of health effects around hazardous waste sites.

The study by Kreiss et al.[139] is an example of a CDC-conducted study. The salient points of this study are summarized, along with other studies, in Table 6. More details of this study are given in Appendix C, which provides a synopsis of the major published and unpublished studies and study protocols reviewed in this section.

The impetus for the Kreiss et al. study was the determination of high DDT residues in fish caught near Triana, Alabama in 1978. Blood samples from twelve residents showed elevated levels of DDT. This cross-sectional study attempted to include blood samples from all residents of Triana and its rural environs. Twenty-eight percent had levels of DDT greater than ten times the U.S. geometric mean although adverse health effects were not associated with DDT.

Rothenberg[140] reported an NIOSH/CDC cross-sectional study around the Hyde Park landfill in New York. The goal was to provide a rapid assessment of health effects. There were limitations that made the study inconclusive. Rather than targeting a specific health effect, it studied 180 variables. Also, there was low participation and insufficient time had elapsed to demonstrate long-term effects such as cancer.

Halperin et al.[141] reported a study conducted following a fire at a waste-chemical disposal plant in New Jersey. There was a possibility of exposure to tetrachlorodibenzofuran (TCDF) or tetrachlorodibenzodioxin (TCDD) of people present at the fire, including fire fighters, police, journalists, and spectators. CDC and the New Jersey Department of Health jointly conducted a cross-sectional health survey of people present at the fire. Soil samples throughout the fire zone and wipe samples of fire equipment were measured for TCDF and TCDD. People at the fire experienced increased respiratory symptoms, but no TCDF or TCDD was detected.

Table 6.   Studies of Adverse Health Effects Resulting from Potential Exposures from Hazardous Waste Sites

Location of Site	Hazardous Materials Involved	Physical Condition	Principal Route of Human Exposure	Population Studied	Study Design	Health Measurements	Exposure Measurements	Results
Triana, Alabama (Kreiss et al.)[139] (CDC)	DDT and related chemicals	Industrial waste dumped into stream	Food chain	Residents of Triana and rural environs	Cross-sectional	Blood analysis, urine analysis, medical questionnaire	Serum DDT residues	Serum triglyceride positively associated with DDT. No association with other health effects.
Hyde Park Landfill, New York (Rothenberg)[140] (NIOSH/ CDC)	Diverse chemicals including chlorinated hydrocarbons	Inactive landfill in residential area	Air	Residents	Cross-sectional	Health questionnaire, urine and blood tests	Urine and blood	9 positive associations out of 180 variables. Hiatus hernia highest association.
Southern New Jersey (Halperin et al.)[141] (CDC/NJ DOH)	Chemicals including PCB, benzene, methylene, chloride, aniline	Fire at disposal plant	Air, direct contact	All people present at fire	Cross-sectional	Questionnaire of respiratory, eye, throat complaints	Soil samples; distance of people from site	Positive association of symptoms with duration of exposure.
Woburn, Massachusetts (Parker et al.) (CDC/MA DOH)[142]	Chemicals include lead, arsenic, organic contaminants	Industrial wastes detected in section of Woburn	Groundwater	Residents of Woburn	Ambidirectional	Cancer incidence	Reported in interview	Greater than expected incidence of childhood leukemia and renal cancer, but no association with environmental hazards.
Superfund Waste Sites (CDC)[31]	PCBs	Not applicable	Not applicable	People exposed to PCBs from Superfund sites	Several stages, including cross-sectional and prospective cohort	Not stated	Serum measurement, questionnaire	Not applicable
Genesee Co., Michigan (Oudbier et al.)[146] (MI DOH)	Diverse chemicals	Inactive incinerator	Air	Residents within 2 square miles around site	Cross-sectional	Questionnaire of respiratory complaints	Soil sampling, length of residence, self-reported exposure to fumes	Male: positive association with respiratory complaints. Female: no association.
Ocean Co., New Jersey (NJ DOH)[147]	Diverse chemicals	Inactive	Groundwater, air	Residents	Cross-sectional	Health questionnaire on skin problems, chronic illness	Household water and air samples	Positive association with skin complaints.

Site	Chemicals	Site type	Exposure route	Population	Study design	Health outcome	Exposure assessment	Results
Camden, Co., New Jersey (NJ DOH)148	Diverse chemicals	Inactive landfill, leachate flowing into creek	Groundwater, air	Residents	Cross-sectional	Questionnaire on eye/throat problems	Household air sampling	Increased respiratory complaints among residents
Price Landfill, New Jersey (NJ DOH)149	Diverse chemicals, including cadmium, lead, chloroform, benzene	Inactive landfill, with erosion	Groundwater	Households affected by groundwater	Cross-sectional	Health questionnaire	Well water sample	Females: increased health complaints, no difference in reported medical problems.
Love Canal, New York (Janerich et al.)11 (NY DOH)	Largely hydrocarbon residue from pesticide production	Inactive landfill	Direct, air, groundwater	Census tract around Love Canal	Ecologic	None	None	No significant increased cancer mortality.
Lorain Co., Ohio (Indian et al.)151 (OH DOH)	Diverse chemicals	Clay lined landfills and lagoon	Air	Residents (northern Eaton township)	Ecologic	None	None	No significant increased mortality.
Clinton County, Pennsylvania (PA DOH)100	Diverse chemicals, including betanaphthylamine, DCB, TCPAA	Storage from chemical company operating until 1981	Air, direct contact	Residents, former employees of chemical company	Cross-sectional and ecologic	Cancer mortality, all morbidity and mortality, urinalysis of former employees	Not stated	Not applicable
Love Canal, New York (Paigen et al.)151 (Independent)	Largely hydrocarbon residue from pesticide production	Inactive landfill	Direct, groundwater, air	Resident children	Ambidirectional	Birth weight, birth height	Distance from site	Significantly lower birth weight and birth height
Utah (Lyon et al.)101 (Independent)	Radon gas	Uranium tailing dump	Air	Residents in two counties containing dump	Ambidirectional	None	None	No significant association between distance from dump and radiogenic malignancies
Canonsburg, Pennsylvania (Talbott et al.)99 (Proposed)98 (Independent)	Radioactive waste	Inactive waste dump next to uranium processing site	Air, direct contact	Adults residing within 1/3 mile of site for at least 15 years (Residents alive in 1938)	Cross-sectional (Retrospective cohort)	Thyroid evaluations, medical history (mortality)	Length of residence	Marginally higher rate of radiation-related thyroid diseases
Hardeman Co., Tennessee (Clark et al.)155 (Meyer)156 (Independent)	Diverse chemicals, including chlorinated organic compounds	Inactive dump, barrels of waste buried in shallow trench	Groundwater, air	Residents	Cross-sectional	Urinalysis for selected organic compounds, blood analysis for liver profile	Household air & water samples	Significant differences in liver profiles; no differences in skin or eye abnormalities

The Massachusetts Department of Public Health (MDPH) reported a study of cancer incidence in Woburn, Massachusetts.[142] Hazardous wastes were discovered in northeast Woburn in 1979, and a large number of childhood leukemia cases were noted by a local clergyman. The MDPH requested the assistance of CDC in further investigation of the health of residents. An ambidirectional study was undertaken in which each case of childhood leukemia, renal cancer, liver cancer, or bladder cancer was matched with two controls. Although comparison with data from the Third National Cancer Survey showed an overall excess of leukemia and renal cancer, the interviews of cases and controls (or their nearest living relatives) showed no association with environmental hazards.

An example of a national study is that of people exposed to PCBs from Superfund sites[31] as previously mentioned. This is a cooperative study by CDC, EPA, and state and local health departments. PCB was chosen as the priority chemical because it is the most prevalent chemical in Superfund sites; it can be reliably measured in serum, and serum levels reflect body burdens over many years.

The study protocol has four parts:

1. Assess Superfund sites for the presence of PCBs and for potential human exposure

2. Conduct pilot studies of potentially "most exposed" people

3. Design and conduct community exposure and health surveys

4. Design and conduct registries of PCB-exposed cohorts detected by community surveys

In addition to EPA, CDC is also involved in an effort to collect and computerize technical data derived from the NPL and other selected sites. Using HRS data and EPA regional site inspection reports, CDC's Agency for Toxic Substances and Disease Registry is independently developing a preliminary databank of information related to 755 hazardous sites including all current NPL sites.[143] Compared to the EPA/Mitre Corporation's NPL technical database, the utility of the CDC file as a resource for health effects evaluations is very limited, however, since chemical exposure information is currently available for only selected sites, and public access to the data is apparently restricted.

Council of Environmental Quality (CEQ)

The CEQ was created by the National Environmental Policy Act of 1969. It is the advisory body in the Executive Office of the President on national policies and programs affecting environmental quality. The CEQ provides some assistance to federal agencies in coordinating programs. It conducts studies on an ad hoc basis; examples are studies on human reproductive hazards from environmental pollution and on groundwater contamination from hazardous waste disposal.

State Government Activities

RCRA was designed to eliminate uncontrolled disposal of hazardous wastes. It lays out a waste management approach with the assumption that states would be delegated the administrative, financing, and enforcement responsibilities of the program. Under RCRA each state is to develop a hazardous waste program for approval by the EPA. State programs must comply with RCRA regulations in order to administer the program without EPA intervention.

Since passage of RCRA, state legislative activities to develop regulatory programs have been intense. Some states have gone beyond mere compliance with RCRA by encouraging industries to investigate other options besides landfilling by providing incentives and disincentives. These can be financial, fee structures, tax incentives, or bonds, to name a few. Legislation can also exclude certain materials or facilities from the regulatory program. Some states have also enacted right-to-know legislation giving workers information about the hazardous materials with which they work. The National Conference of State Legislatures documented the hazardous waste legislation of 28 states in *Hazardous Waste Management: A Survey of State Legislation, 1982.*[144]

After passage of RCRA, there was still concern about abandoned or inactive sites. In 1980 Congress passed CERCLA to help pay for remedying abandoned sites. Usually the responsible party pays for the cleanup, but if that party cannot be found or is not financially able to clean up the site, then the state and federal governments bear the financial burden. At CERCLA sites, the costs of cleanups paid for by the states range from 10% at privately owned sites to 50% at publicly owned sites. States are also responsible for sites not addressed by CERCLA. To prevent similar problems with abandoned sites from developing in the future, states have developed comprehensive procedures for siting new facilities.

Between October 1982 and January 1983, the New Mexico Health and Environmental Department surveyed the 50 states to determine which states had environmental epidemiology programs.[145] The state epidemiologists from the 26 states which responded were queried about epidemiologic studies of populations with exposure to hazardous waste sites. Their responses are summarized positively in Table 7. Seventeen of the 26 states have completed health studies at some time in the past. Eleven now have ongoing or planned studies. Only three states have active registries of persons residing near hazardous waste sites. Examples of state-conducted studies, which are summarized in Table 6 and Appendix C, are the study by Oudbier et al.,[146]

Table 7. State-Conducted Health Studies Around Hazardous Waste Sites

State	Completed Studies	Ongoing or Planned Studies	Active Registry of Exposed Population
Kansas	+	+	+
New York	+[a]	+	+
California	+	+	0
Colorado	+	+	0
Connecticut	+	+	0
Massachusetts	+[b]	+	0
Michigan	+[c]	+	0
Ohio	+[d]	+	0
Pennsylvania	+	+[f]	0
New Jersey	+[e]	0	+
Louisiana	+	0	0
Maine	+	0	0
Maryland	+	0	0
New Mexico	+	0	0
Texas	+	0	0
Utah	+	0	0
Virginia	+	0	0
Idaho	0	+	0
Washington	0	+	0
Illinois	0	0	0
North Carolina	0	0	0
Oklahoma	0	0	0
Rhode Island	0	0	0
Wisconsin	0	0	0
Arizona	0	0	0
South Dakota	0	0	0

[a]From Janerich et al.[11] and Rothenberg, R.[140]
[b]From Parker, G. S. and Rosen, S. L.[142]
[c]From Oudbier et al.[146]
[d]From Indian et al.[150]
[e]From Halperin et al.,[141] Conomos, M. G.,[147] and the New Jersey Department of Health.[148-149]
[f]From the Pennsylvania Department of Health.[100]

the three studies by the New Jersey Department of Health,[147-149] the studies by Indian et al.,[150] and by Janerich et al.[11]

State health departments have a responsibility to respond to citizen complaints. The first four studies mentioned were cross-sectional questionnaire surveys. All of them had some positive response regarding either respiratory or skin complaints. However, a potential source of bias exists when studies are initiated by resident complaints. In this situation, study subjects are well aware of the purpose of the study, and their perceived health status may be affected by positive study results. It is difficult to obtain objective answers from a questionnaire under these conditions, particularly if the outcome variable is a subjective reporting of relatively short-term complaints.

Establishing the potential for exposure before launching an epidemiologic investigation is the desirable approach. All four studies made some attempt to determine the extent or possibility of exposure. Oudbier et al.[146] collected soil samples from 49 sites. The study in Ocean County, New Jersey took air samples in selected households; an earlier investigation had demonstrated contamination of the aquifer with a variety of organic chemicals.[147] The Camden County study gave results of air sampling both inside and outside homes.[148] Contamination of private and public wells had been established by the EPA, DEP, and Atlantic County Health Department before the Price Landfill study was undertaken.[149]

Unlike the cross-sectional surveys discussed above, the study by Indian et al.[150] in Lorain County, Ohio was an ecologic study of cancer mortality rates. It was not conclusive because of the inability to relate county cancer mortality to individual exposure. Also, there were many confounding variables; race, age, and geographic location were accounted for but socioeconomic status, diet, smoking, occupation, and familial history of cancer were not. Finally, small sample sizes (relatively few cancer deaths in the county) gave the study little statistical power.

The previously mentioned planned study of the Drake Chemical site proposed by the Pennsylvania Department of Health[100] is one of the few that will attempt to define a cohort retrospectively and follow their health outcomes through time. Cross-sectional evaluation of health and an ecologic examination of cancer mortality are also planned.

The studies of Love Canal by Janerich et al.[11] and Paigen and Goldman[151] are two of the many studies at Love Canal. Reviews of investigations of the Love Canal incident are given by Kimbrough et al.[32] and Heath.[152] The problems at Love Canal occurred from 1942 to 1953 when a 3000-foot long ditch was filled with chemical wastes arising from the manufacturing of pesticides. In 1953 it was sealed with a clay cap. Over the next 10 to 15 years the surrounding area was developed as a residential neighborhood. In the 1970s it became apparent that chemicals were leaking through a broken cap. Homes closest to the site were evacuated, and the site was resealed with clay.

The first assessments of health were in 1978 by the New York State Department of Health (NYSDH).[10] They conducted environmental testing for chemicals in soil, air, water, and in the homes. There was also a health questionnaire of nearby residents.

In 1979 additional testing was carried out by NYSDH and EPA. Studies were also begun by community groups and groups such as the Environmental Defense Fund. A pilot study was done in 1980 under EPA auspices on possible chromosome aberrations in Love Canal residents.[153] It showed chromosomal damage in 11 of 36 people tested for breakage, and it was partially responsible for the relocation of families from the area. However, much scientific controversy was raised over the study's methodology and interpretation of results, due largely to the apparently inadequate control material.[154]

In 1981 Janerich et al.[11] compared cancer incidence, as ascertained by the

New York Cancer Registry, at Love Canal to the entire state outside of New York City. There was no evidence of any increased cancer associated with toxic wastes buried at the dump site. As with other ecologic studies of dump sites, this study had the limitation of a small study population. Also, since cancers were the outcome variable, there are uncertainties related to the latent periods of the diseases.

The study by Paigen and Goldman[151] was one of the investigations initiated by local residents who eventually had grown to distrust government at all levels. This study reported an association of low birth weight and low birth height among children of Love Canal homeowners. A drawback of this and other health studies, however, is the possibility of misclassification bias. Individuals' exposures to chemicals were simply not known, and they cannot be determined because most of the chemicals are not biologically persistent.

*Really don't
true tric*

Independent Research

The studies by Paigen and Goldman,[151] Lyon et al.,[101] Talbott et al.,[98] Clark et al.,[155] and Meyer[156] represent independent research. With a relative absence of political pressure, independent researchers have the opportunity to investigate different epidemiologic methods. Paigen and Goldman,[151] as discussed above, used birth weight and birth height as health measures. Lyon et al.[101] utilized an ambidirectional design to study clustering of cases drawn from the Utah Cancer Registry with environmental hazards. Talbott et al.[98] is investigating methods for developing a retrospective cohort of persons living near a uranium waste site.

Lyon et al.[101] drew cases and controls from 1966 to 1975 from the Utah Cancer Registry. Addresses were determined and plotted on maps. The distance to the site was used as a measure of exposure. This technique would be useful if distance from the site is an appropriate surrogate measure of exposure. In his study this is questionable since prevailing wind direction was not taken into account.

Talbott et al.[98] conducted a cross-sectional study of residents living near a uranium waste site compared to appropriate controls. There was a slight increase of radiation-related thyroid abnormalities. A followup is being planned that would define and trace a retrospective cohort.[98]

Clark et al.[155] and Meyer[156] studied the waste site in Hardeman County, Tennessee. Over 300,000 55-gallon drums of solid and liquid pesticide wastes were buried in shallow trenches between 1964 and 1972 before the state ordered the dumping stopped. In the 1970s it became apparent that well water had been contaminated by the site. Most residents had stopped using well water for potable supply by 1978; all uses of the well water ceased in 1979.

Clark et al.[155] initiated a cross-sectional health study in 1978; it had to be conducted quickly before much time had elapsed after use of well water had stopped. The population was divided into three groups—high exposure, intermediate exposure, unexposed—depending on the amount of carbon tetrachloride in their well water. The survey used a health questionnaire, a clinical exam, and a biochemical screening that included serum analyses for liver and kidney function parameters. There was an elevated liver profile in residents using the water, but no skin or eye abnormalities. Results were inconclusive, partially due to the delay in starting the study and the necessary haste in completing it. In their conclusion, Clark et al.[155] make a plea for better anticipation of problem situations.

There are two key features of the published studies examined in Table 6 that are undoubtedly a consequence of the numerous methodologic difficulties described in earlier sections.

First, in view of the many thousands of hazardous waste sites that have been identified as posing potential health problems to large human populations, it is remarkable that, to date, only about 15 published health effects evaluations (involving 13 different sites) were identified following a fairly extensive review of the scientific literature.

The second feature is evident upon examination of the health measurements and results columns of Table 6. Much of the health outcome data in many of the studies were derived from questionnaires which are generally subjective and nonspecific. Moreover, due to one or more drawbacks including small sample size, lack of exposure or toxicologic data, and the inability to control for confounding factors, the exposure-health outcome associations that were examined are, for the most part, weak or inconclusive.

In summary, the overall paucity of published research coupled with the largely inconclusive results that have been produced to date again underscore the need and importance of developing research programs that will lead to a more complete understanding of the health problems derived from hazardous waste sites.

CONCLUSIONS AND RECOMMENDATIONS

In order to provide a basis and impetus for research related to the assessment of human health effects from exposure to hazardous waste sites, this chapter has attempted to describe the current and projected scope of the waste site problem, appraise the classical and propose alternative epidemiologic approaches to health evaluations, review completed ongoing and proposed research activities and programs in this area, and assess the need for and the nature of future research. This section summarizes the major findings and conclusions of this extensive review and recommends several general and specific areas of research that may lead to a more complete understanding and more valid assessment of current and future waste site-related health problems.

Major Findings and Conclusions

1. From a public health standpoint, the contamination of drinking water probably represents the largest waste site-related health problem since EPA has estimated that as many as 30,000 waste sites may pose significant health problems related to ground water contamination and over 40% of the U.S. population currently relies on ground water as a source of drinking water.

2. The hazardous waste site problem that exists today is, curiously, an unpredicted effect of the strong water and air sanctions enacted between 1952 and 1977 in which the water and air were no longer considered acceptable sinks for the disposal of wastes, leading industries and municipalities to increasingly turn to the land for waste disposal.

3. The most significant hazardous and solid waste laws that have been passed to date are The Resource Conservation and Recovery Act (RCRA) of 1976 and the Comprehensive Environmental Response, Compensation and Liability Act (CERCLA) of 1980. RCRA provides regulation of currently operating hazardous waste sites and for the first time provided the federal EPA with enforcement power. CERCLA, also known as "Superfund," was enacted as a five-year plan to provide federal dollars and authority to respond to emergencies and to take remedial action at abandoned hazardous waste sites.

4. By the middle of 1983, EPA had inventoried almost 16,000 uncontrolled hazardous waste sites of which the worst 539 sites appeared on the August 1983 National Priority List. Selection of these sites was based primarily on scores derived from the EPA hazard ranking system developed in 1982 by the Mitre Corporation.

5. While it is speculated that site assessments and inspections, remedial action and other work are currently performed at a rate that should successfully address priority sites within the current framework of CERCLA, recent reports by the Office of Technology Assessment and the National Academy of Sciences have warned that the process of cleaning up the Superfund sites may itself lead to the creation of new Superfund sites. The OTA and NAS reports emphasize the critical need for continued research in the alternative forms of waste disposal (incineration, thermal methods, recycling, biological degradation) in order to avert the perpetuation of hazardous waste sites.

6. The classical application of epidemiology to assess relationships between toxic waste site exposure and possible health consequences is made difficult not only by the limitations inherent in all observational studies of human populations but also by the number of particularly complex real-life situations which uniquely characterize hazardous waste site studies. These

difficulties include: populations living in the vicinity of hazardous waste sites are usually small, limiting both the range of outcomes and the size of the effects that can be studied; the persons living in any given area are usually heterogeneous with respect to characteristics that can influence many health outcomes independently of exposure or with respect to the type, level, duration, or timing of the exposure; there is generally a lack of exposure information and of toxicologic data related to mixtures or to combinations of chemicals; health effects of exposure to chemicals depend on the nature of the dump materials, the integrity of containment, and underlying hydrogeology and surrounding geography; many of the health end points of interest are either rare, are associated with long variable latency periods, or are unlikely to have been routinely recorded prior to investigations; a general lack of specificity of common clinical indicators of disease; the instruments used to measure health outcomes are generally very insensitive; publicity related to episodes under study may produce or accentuate reporting bias; and the conduct of waste site studies is made difficult due to the presence of a highly charged atmosphere of anger and fear which often accompany suspicion of adverse health effects.

7. Since 1981 at least four large meetings were convened to consider the public health and technical problems associated with hazardous waste sites and to evaluate approaches to both monitoring exposures to toxic substances and to linking these exposures to possible adverse health effects. The general consensus of these meetings was that reproductive alterations, chromosomal damage, and neurotoxic effects were the most promising health end points for epidemiologic investigations of waste site-exposed populations. Other generally less feasible health end points include effects on tissues and organ function, cancer incidence, and mortality.

8. The classical epidemiologic study designs that can be considered to evaluate exposure-health outcome relationships at hazardous waste sites include the basic designs (cohort, cross-sectional, and case control), the hybrid designs (combining or extending elements from the basic designs), and incomplete designs (where information is missing on one or more relevant factors, e.g., ecologic studies and space, time, and cluster designs). In addition, the establishment of the exposure-specific or waste site-specific registries of exposed persons may be useful to provide crucial information needed by state and federal agencies who may recognize the need for epidemiologic studies for research or for determining who might require health care and long-term followup.

9. Since the classic task of epidemiology to evaluate health effects at hazardous waste sites is made difficult due to a myriad of methodologic limitations and particularly complex real-life situations, it is crucial that environmental epidemiologists begin to develop methods to enhance the analytic capabilities of these approaches and to consider alternative approaches. This chapter proposes four categories of nonclassical

approaches that might be pursued: to increase the inferential capabilities of existing methods by increasing analytic control over extraneous factors or by decreasing the dependency of the methods on underlying assumptions or requirements; to explore familiar roles for epidemiology in nonenvironmental settings and to utilize these roles as paradigms for possible roles in hazardous waste site settings; to consider other nonepidemiologic methods that are used to assess analogous problems in nonenvironmental settings; and to employ the classical methods of epidemiology to study the health experience of occupational groups that are heavily exposed to substances commonly found in waste sites.

10. Several large-scale databases were identified that may be utilized in epidemiologic investigations at hazardous waste sites. For example, the National Priority List databases at the Centers for Disease Control in Atlanta and at the Mitre Corporation in McLean, Virginia are national level data repositories that may provide useful summary data on potential exposures specific to geographic areas that contain toxic waste sites. Some of the larger sources of data on morbidity and mortality that may be useful in designing and conducting economic analyses include: the National Center for Health Statistics that publishes summary data on births and deaths at the national, state, and metropolitan level; the University of Pittsburgh Graduate School of Public Health and the Johns Hopkins School of Hygiene and Public Health, who have independently developed computer systems capable of generating factor-specific death rates at the state or county level; the National Cancer Institute's Surveillance, Epidemiology, and End Results program that provides cancer incidence data for about 10% of the U.S. population residing in the major standard metropolitan statistical areas; the SEEDIS and UPGRADE computer systems that integrate mortality and morbidity statistics with various sources of environmental data; and the chemical substances information network developed jointly by the EPA and the Council on Environmental Quality in accordance with the Toxic Substances Control Act that links together many computer data systems providing information on chemicals toxicity production and environmental effects.

11. Waste site-related responsibilities in research activities at the federal level involve primarily the Environmental Protection Agency (EPA) and the Centers for Disease Control (CDC). The EPA's role is basically environmental control while CDC is more directly involved with the health of persons residing at hazardous waste sites. Regarding state level activities, a recent survey conducted by the New Mexico Health and Environmental Department revealed that 26 states currently have active environmental epidemiology programs. Seventeen of the 26 states have completed health studies some time in the past, and 11 now have ongoing or planned studies. Only three states currently have active registries of persons residing near hazardous waste sites.

12. A systematic literature review utilizing *Index Medicus* and two computer database indices, *Enviroline* and *Toxline*, revealed only 15 published health effects evaluations at hazardous waste sites (involving only 13 different sites). Most of these studies were cross-sectional, deriving exposure and health outcome data from generally subjective and nonspecific questionnaires. Moreover, due to one or more methodologic drawbacks, the exposure-health outcome associations that were examined are, for the most part, weak or inconclusive.

13. Some of the major ongoing or proposed health effects evaluations at hazardous waste sites include: CDC's multistage study of persons exposed to PCBs at Superfund waste sites; the Pennsylvania State Health Department's proposed retrospective cohort study of former employees and selected residents of Lock Haven, Pennsylvania, who may have experienced hazardous exposures associated with the Drake Chemical Superfund site; and the University of Pittsburgh Center for Environmental Epidemiology proposed retrospective cohort study of a community exposed since 1910 to low-level radon and gamma radiation.

Recommendations

The general and specific recommendations in this chapter are grouped according to the three fundamental phases of health effects evaluations that were described above: exposure assessment and definition of exposed and unexposed populations, measurement of health outcomes, and determination of exposure-health outcome relationships.

Exposure Assessment and Definition of Exposed and Unexposed Populations

1. Determine the toxicity of waste site materials giving consideration to mixtures and the effect of storage or movement in water or the ground.

2. Develop centralized databanks containing toxicological data on chemicals that are likely to be found at waste sites. In these databanks, toxicology data gleaned from waste site studies should be integrated with information from the literature in readily retrievable form to facilitate its use in the public and private sectors.

3. Develop techniques, such as those used in occupational studies, for the passive, noninvasive, and quantitative monitoring of individual exposures in community populations.

4. Determine baseline exposure levels in the general population of chemicals likely to be found at waste sites.

5. Because the frequent need to analyze multiple samples of air, soil, and water for many chemicals can often overwhelm laboratory facilities, priorities need to be developed to limit the number of chemicals to be studied and to minimize the required number of samples.

6. Establish national reference laboratories for exposure and biological monitoring data. There is also a need for more work on monitoring inter-laboratory comparability among those laboratories participating in waste site exposure evaluations.

7. Identify and categorize the major dump sites by source, type, and potential for health risks. This effort could be facilitated by the development of site hazard ranking schemes that place more emphasis on the potential for adverse health effects. Regarding this research need, CDC recommends that a set of decision criteria be mutually developed by EPA and the Department of Health and Human Resources to establish the degree of human health hazard at particular waste sites. This might include additional monitoring and chemical analysis to quantify the chemicals and to determine the degree of exposure. It might also include health examinations to assess whether an exposed population is experiencing disease or adverse health effects.

Measurement of Health Outcomes

1. Develop survey techniques and survey instruments that are resistant to the psychological and emotional issues involved in hazardous waste site study settings. This entails the increased development and implementation of measurements involving verifiable events, such as laboratory procedures and records, hospital or physician records, absenteeism, and factual information on morbid or vital data.

2. Develop tests to preclinically recognize the effects of chemicals in human populations. This includes: tests for preclinical stages of cancer, reproductive disorders, cytogenetic disorders, and diseases of specific organ systems such as the central nervous system, liver, and kidney.

3. Increase efforts by the NCHS to generate more current and demographically detailed data on total and cause-specific mortality. Emphasis should be placed on the sentinel diseases, for example, diseases which are relatively rare and that may be associated with environmental exposures.

4. Further develop state and local registries of morbid events such as malignant diseases and congenital malformations to facilitate the conduct of epidemiologic studies.

Determination of Exposure-Health Outcome Relationships

1. Develop and apply alternative methods to the classical epidemiologic analytic approaches to health effects evaluations. There are four general categories of alternative approaches that might be pursued: (1) increase the inferential capabilities of existing statistical and epidemiologic methods by increasing analytic control over extraneous factors (e.g., multivariate methods), or by decreasing the dependency of the methods to underlying

assumptions or requirements (e.g., nonparametric alternatives); (2) explore familiar roles for epidemiology in nonenvironmental settings and to utilize these roles as paradigms for possible roles in hazardous waste settings; (3) consider other nonepidemiologic methods that are used to assess analogous problems in nonenvironmental settings; and (4) employ the classical methods of epidemiology to study the health experience of occupational groups that are heavily exposed to substances commonly found in waste sites.

2. Plan and implement historical-prospective studies of cohorts living near toxic waste sites. Efforts should be made to identify cohorts, historically, beginning at the time the waste site was established. Provisions should be incorporated for future linkage of these data sets with exposure estimates and disease registries.

3. Develop exposure-specific and site-specific registries of exposed populations and suitable control populations with provisions for short-term monitoring and long-term followup.

4. Further develop automated data systems of births, deaths, and diseases, such as the National Death Index, that are important to the success of long-term followup studies.

REFERENCES

1. Wilkes, A., Kiefer, I., and Levine, B. Everybody's Problem: Hazardous Waste. U.S. EPA, Office of Solid Waste, SW-826, 1980.
2. Report of the Subcommittee on the Potential Health Effects of Toxic Chemical Dumps of the DHEW Committee to Coordinate Environmental and Related Problems. Department of Health, Education, and Welfare, May, 1980.
3. Resource Conservation and Recovery Act (RCRA). PL No. 94-580, 90 Stat. 2795, 1976.
4. Toxic Substances Control Act (TSCA). PL No. 94-469, 90 Stat. 2003, 1976.
5. Hart, F. C. Preliminary Assessment of Cleanup Costs of National Hazardous Waste Problems. Report to the U.S. EPA, Washington, DC: EPA, February 23, 1979.
6. Landrigan, P. J., and Gross, R. L. Chemical wastes—illegal hazards and legal remedies. Editorial. *Am. J. Public Health* 71:985–987, 1981.
7. Singal, M., Kominsky, J. R., Schulte, P. A., and Landrigan, P. J. Technical Assistance Final Report No. 79-022-789: Hyde Park Landfill, Niagara, NY. Cincinnati, OH: National Institute for Occupational Safety and Health, 1980.
8. Trichloroethylene exposure—Pennsylvania. *Morbidity Mortality Weekly Report* 30:226–233, May 22, 1981.
9. Baker, E. L., Field, P. H., Basteyns, B. J., Skinner, G. H., et al. Phenol poisoning due to contaminated drinking water. *Arch. Environ. Health* 33:89–94, 1978.
10. Department of Health, State of New York. Love Canal, Public Health

Time Bomb. A Special Report to the Governor and Legislature. New York: September, 1978.

11. Janerich, D. T., Burnett, W. S., Feck, G., Hoff, M., et al. Cancer incidence in the Love Canal area. *Science* 212:1404–1407, 1981.

12. Miller, D. W. Geohydrological surveys at chemical disposal sites. In Lowrance W. W. (Ed.): *Assessment of Health Effects at Chemical Disposal Sites.* Proceedings of a symposium at Rockefeller University, June 1–2, 1981. New York: Rockefeller University, 1981.

13. Lowrance, W. W. Interpretive summary of the symposium. In Lowrance W. W. (Ed.): *Assessment of Health Effects at Chemical Disposal Sites.* Proceedings of a symposium at Rockefeller University, June 1–2, 1981. New York: Rockefeller University, 1981.

14. Tarr, J. A. Chapter 2, this volume.

15. Comprehensive Environmental Response, Compensation and Liability Act (CERCLA). Public Law 96-510. 94 Stat 2767, December 11, 1980.

16. Goldstein P. Status and assessment of chemical waste sites. In *Evaluations of Health Effects from Waste Disposal Sites.* Presented at the Fourth Annual Symposium on Environmental Epidemiology at the University of Pittsburgh Graduate School of Public Health, May 2–4, 1983.

17. National Oil and Hazardous Substances Contingency Plan. *Federal Register* 47 (137):31180, July 16, 1982 (to be codified at 40 CSR. § 300).

18. Hazardous Waste Sites, National Priority List. Washington, DC: U.S. EPA Office of Solid Waste and Emergency Response, August, 1983.

19. Technologies and Management Strategies for Hazardous Waste Control, Congress of the United States. Washington, DC: Office of Technology Assessment, 1983.

20. Management of Hazardous Industrial Wastes. Research and Development Needs. National Materials Advisory Board, Commission on Engineering and Technical Systems, National Research Council. Washington, DC: National Academy Press, 1983.

21. Hileman, B. Hazardous waste control: Are we creating problems for future generations? *Environ. Sci. Technol.* 17:281–285, 1983.

22. Landrigan, P. J. Epidemiologic approaches to persons with exposures to waste chemicals. *Environ. Health Pers.* 48:93–97, 1983.

23. Heath, Jr., C. W. Field epidemiologic studies of populations exposed to waste dumps. *Environ. Health Pers.* 48:3–7, 1983.

24. Maugh, T. H., II. Just how hazardous are dumps? *Science* 215:490–493, 1982.

25. Goyer, R. A. Introduction and overview. *Environ. Health Pers.* 48:1–2, 1983.

26. Tidwell, D. C., Brown, G. E., Jr., and Sidney, S. Chemical Substances Information Network (CSIN): An Overview. Interagency Toxic Substances Data Committee, 1983.

27. Maugh, T. H., II. Biological markers for chemical exposure. *Science* 215: 643–647, 1982.

28. Stein, Z., Hatch, M., Kline, J., Shrout, P., et al. Epidemiologic considerations in assessing health effects at toxic waste sites. In Lowrance, W. W. (Ed.): *Assessment of Health Effects at Chemical Disposal Sites*. Proceedings of a symposium at Rockefeller University, June 1–2, 1981. New York: Rockefeller University, 1981.
29. Clarkson, T. W., Weiss, B., and Cox, C. Public health consequences of heavy metals in dump sites. *Environ. Health Pers*. 48:113–127, 1983.
30. Health and Nutrition Surveys, National Center for Health Statistics. 3700 East-West Highway, Hyattsville, MD 20782.
31. Centers for Disease Control, Center for Environmental Health, Chronic Diseases Division and Clinical Chemistry Division. Polychlorinated Biphenyls (PCB) Exposure at Superfund Wastesites. Unpublished study protocol, August, 1983.
32. Kimbrough, R. D., Taylor, P. R. Zach, M. M., and Heath, C. W. Studies of human populations exposed to environmental chemicals: Considerations of Love Canal. In: *Assessment of Multichemical Contamination*, Proceedings of an International Workshop, Milan, Italy, April 28–30, 1981. Washington DC: National Academy Press, 1982, pp. 289–306.
33. Report of Panel II. Guidelines for reproductive studies in exposed human populations. In Bloom, A. D. (Ed.): *Guidelines for Studies of Human Populations Exposed to Mutagen and Reproductive Hazards*. White Plains, NY: March of Dimes Birth Defects Foundation, 1981.
34. Corn, M., Emmett, E. A., Breysse, P. N., and Gold, E. B. Design of Environmental Health Effects Studies Stemming from the Potential for Human Exposure to Toxic Waste in Memphis, Tennessee. Unpublished Final Report. The Johns Hopkins University School of Hygiene and Public Health, Baltimore, MD, December 7, 1981.
35. Schaumberg, H. H., Arezzo, J. C., Markowitz, L., and Spencer, P. S. Neurotoxicity assessment at chemical disposal sites. In Lowrance, W. (Ed.): *Assessment of Health Effects at Chemical Disposal Sites*. Proceedings of a symposium at Rockefeller University, June 1–2, 1981. New York: Rockefeller University, 1981.
36. Bloom, A. D. (Ed.). *Guidelines for Studies of Human Populations Exposed to Mutagenic and Reproductive Hazards*. White Plains, NY: March of Dimes Birth Defects Foundation, 1981.
37. Lowrance, W. W. (Ed.). *Assessment of Health Effects at Chemical Disposal Sites*. Proceedings of a symposium at Rockefeller University, June 1–2, 1981. New York: Rockefeller University, 1981.
38. National Institute of Environmental Health Sciences. Health Effects of Toxic Wastes. *Environ. Health Pers*., Vol. 48, February, 1983.
39. Warburton, D. Selection of human reproductive effects for study. In Lowrance, W. W. (Ed.). *Assessment of Health Effects at Chemical Disposal Sites*. Proceedings of a symposium at Rockefeller University; June 1–2, 1981. New York: Rockefeller University, 1981.
40. Wharton, D., Kraus, R. M., Marshall, S., and Milby, T. Infertility in male pesticide workers. *Lancet* 2:1259–1261, 1977.

41. Taylor, J. R., Selhorst, J. B., and Colabrase, V. P. Chlordecone. In Spender, P. S. and Schaumberg, H. H. (Eds.): *Experimental and Clinical Neurotoxicology*. Baltimore, MD: Williams & Wilkins, 1980, p. 407.
42. Wyrobeck, A. J. Methods for evaluating the effects of environmental chemicals on human sperm production. *Environ. Health Pers.* 48:53–59, 1983.
43. Wyrobeck, A. J., Watchmaker, G., Gordon, L., Wang, K., et al. Sperm shape abnormalities in carbaryl-exposed employees. *Environ. Health Pers.* 40:255–265, 1981.
44. Knill-Jones, R. R. Rodrigues, L. V., Moir, D. D., and Spence, A. A. Anesthetic practice and pregnancy; controlled survey of women anesthetists in the United Kingdom. *Lancet* 1:1326–1328, 1972.
45. Knill-Jones, R. R., Newman, B. J., Spence, A. A. Anesthetic practice and pregnancy. *Lancet* 2:807–809, 1975.
46. American Society of Anesthesiologists. Report of the ad hoc committee on the effect of trace anesthetics on the health of operating room personnel. *Anesthesiology* 41:321–340, 1974.
47. Infante, P. F., Wagoner, J. K., McMichael, A. J., and Falk, H. Genetic risks of vinyl chloride. *Lancet* 1:734–735, 1976.
48. U.S. Environmental Protection Agency. Report of an assessment of a field investigation of six-year spontaneous abortion rates in three Oregon areas in relation to forest 2,4,5-T practices. Washington, DC: U.S. EPA, 1979.
49. Vianna, N. Adverse pregnancy outcomes—potential endpoint of human toxicity in the Love Canal, preliminary results. In Porter, I. H. and Hook, E. B. (Eds.): *Human Embryonic and Fetal Death*. New York: Academic Press, 1980, pp. 165–168.
50. Nordstrom, S., Beckman, L., and Nordenson, I. Occupational and environmental risks in and around a smelter in northern Sweden. V. Spontaneous abortion among female employees and decreased birthweight in their offspring. *Hereditas* 90:291–296, 1979.
51. Wolff, S. Problems and perspectives in the utilization of cytogenetics to estimate exposure at toxic chemical waste dumps. *Environ. Health Pers.* 48:25–27, 1983.
52. Wolff, S. Cytogenetic analyses at chemical disposal sites: problems and prospect. In Lowrance, W. W. (Ed.): *Assessment of Health Effects at Chemical Disposal Sites*. Proceedings of a symposium at Rockefeller University, June 1–2, 1981. New York: Rockefeller University, 1981.
53. Evans, H. J., Buckton, K. E., Hamilton, G. E., and Carothers, A. Radiation induced chromosome aberrations in nuclear-dockyard workers. *Nature* 277:531–544, 1979.
54. Funes-Cravioto, F., Kolmodin-Hedman, B., Lindsten, J., and Nordenskjold, M. et al. Chromosome aberrations and sister chromatid exchange in workers in chemical laboratories and a rotoprinting factory and in children of women laboratory workers. *Lancet* 2(8033):322–325, August, 1977.

55. Waksvik, H., Klepp, O., Brogger, A. Chromosome analyses of nurses handling cytostatic agents. *Cancer Treatment Reports* 65(7–8):607–610, 1981.
56. Garry, P. F., Hozier, J., Jacobs, D., Wade, R. L., et al. Ethylene oxide: Evidence of human chromosomal effects. *Environ. Mutagen.* 1:375–382, 1979.
57. Schaumberg, H. H., Spencer, P. S., and Arezzo, J. C. Monitoring potential neurotoxic effects of hazardous waste disposal. *Environ. Health Pers.* 48:61–64, 1983.
58. Schaumberg, H. H., Arezzo, J. C., Markowitz, L., and Spencer, P. S. Neurotoxicity assessment at chemical disposal sites. In Lowrance, W. W. (Ed.): *Assessment of Health Effects at Chemical Disposal Sites.* Proceedings of a symposium at Rockefeller University, June 1–2, 1981. New York: Rockefeller University, 1981.
59. Schaumberg, H. H., and Spencer, P. S. The toxic neuropathies – a review. *Neurology* 29:429–431, 1979.
60. Buchthal, F. and Belise, F. Electrophysiology and nerve biopsy in men exposed to lead. *Brit. J. Ind. Med.* 36:135–143, 1979.
61. Seppalainen, A. M., Raitta, C., and Huuskonen, M. S. n-Hexane induced changes in visual evoked potentials and electroretinograms of industrial workers. *Electroenceph. Clin. Neurophysiol.* 47:492–498, 1979.
62. Seppalainen, A. M., Savolainen, K., and Kovala, T. Changes induced by xylene and alcohol in human evoked potentials. *Electroenceph. Clin. Neurophysiol.* 51:148–158, 1981.
63. Seppalainen, A. M. Diagnostic utility of neuroelectric measures in environmental and occupational medicine. In Otto, D. A. (Ed.): *Multidisciplinary Perspectives in Event-Related Brain Potential Research.* Washington, DC: U.S. Government Printing Office, 1978, pp. 448–452.
64. Arezzo, J. C., and Schaumberg, H. H. The use of the Optacon as a screening device. *J. Occup. Med.* 22:461–469, 1980.
65. Arezzo, J. C., Schaumberg, H. H., and Peterson, C. Rapid screening for peripheral neuropathy: A field study with the Optacon. *Neurology*, in press.
66. Lindstrom, K. Changes in psychological performance of solvent-poisoned and solvent-exposed workers. *Am. J. Ind. Med.* 1:69–78, 1980.
67. Seppalainen, A. M., Lindstrom, K., and Martelin, T. Neurophysiological and psychological picture of solvent poisoning. *Am. J. Ind. Med.* 1:31–39, 1980.
68. Brown, G. G., Preisman, R. C. Anderson, M. D., Nixon, R. K., et al. Memory performance of chemical workers exposed to polybrominated biphenyls. *Science* 212:1414–1416, 1981.
69. Kimbrough, R. D. Methodology of clinical studies of exposed populations. In Tucker, R. R., Young, A. L., Gray, A. P. (Eds.): *Human and Environmental Risks of Chlorinated Dioxins and Related Compounds.* Plenum Publishing Corporation, 1983.

70. Meyer, C. R. Liver dysfunction in residents exposed to leachate from a toxic waste dump. *Environ. Health Pers.* 48:9-13, 1983.
71. Rifkind, B. M. The lipid research clinic's population studies data book. Vol. 1. The prevalence study. U.S. DHHS, NIH Publication No. 80-1527, 1980.
72. Kuller, L. H. Approaches to Evaluating Health Effects at Hazardous Wastesites. Presented at the Fourth Annual Symposium on Environmental Epidemiology, University of Pittsburgh, Pittsburgh, PA, May 2-4, 1983.
73. Miller, R. G. *Simultaneous Statistical Inference.* New York: McGraw-Hill, 1966.
74. Fleiss, J. *Statistical Methods for Rates and Proportions,* Second ed. New York: John Wiley & Sons, 1981.
75. Schlesselman, J. J. *Case Control Studies. Design, Conduct, Analysis.* New York: Oxford University Press, 1982.
76. Sackett, D. L. Bias in analytic research. *J. Chronic Dis.* 32:51-63, 1979.
77. Miettinen, O. S. Principles of epidemiologic research. Unpublished manuscript. Cambridge, MA: Harvard University, 1976.
78. Kleinbaum, D. G., Kuller, L. L., and Morgenstern, H. *Epidemiologic Research: Principles and Quantitative Methods.* Belmont, CA: Lifetime Learning Publications, 1982.
79. Breslow, N. E. and Day, N. E. *Statistical Methods in Cancer Research. Vol. 1 – the Analysis of Case-Control Studies.* International Agency for Research on Cancer. Publication No. 32. Lyon, 1980.
80. Anderson, S., Aquier, A., Hauck, W. W., Oakes, D., et al. *Statistical Methods for Comparative Studies.* New York: John Wiley & Sons, 1980.
81. Koopman, J. S. Causal models and sources of interaction. *Am. J. Epidemiol.* 106:439-444, 1977.
82. Kupper, L. L., and Hogan, M. D. Interaction in epidemiologic studies. *Am. J. Epidemiol.* 108:447-453, 1978.
83. Rothman, K. J. Synergy and antagonism in cause effect relationships. *Am. J. Epidemiol.* 99:385-388, 1974.
84. Rothman, K. J. Occam's razor pares the choice among statistical models. *Am. J. Epidemiol.* 108:347-349, 1978.
85. Walter, S. C., and Holford, T. R. Additive, multiplicative, and other models for disease risks. *Am. J. Epidemiol.* 208:314-346, 1978.
86. National Death Index. User's Manual, U.S. Department of Health and Human Services, National Center for Health Statistics, Hyattsville, MD, September, 1981.
87. McMahon, B., and Pugh, T. F. *Epidemiology: Principles and Methods.* Boston: Little, Brown, 1970.
88. Friedman, G. D. *Primer of Epidemiology.* Second ed. New York: McGraw-Hill, 1980.
89. Case, R. A. Cohort analysis of mortality rates as an historical or narrative technique. *Brit. J. Prev. Soc. Med.* 10:159-171, 1956.

90. Lloyd, J. W. and Ciocco, A. Long-term mortality study of steelworkers. I. Methodology. *J. Occup. Med.* 11:299–310, 1969.
91. Enterline, P. E. and Marsh, G. M. Cancer among workers exposed to arsenic and other substances in a copper smelter. *Am. J. Epidemiol.* 116:895–911, 1982.
92. Cooper, W. C., Enterline, P. E., and Warden, E. T. Estimating occupational disease hazards through medical care plans. *Public Health Reports* 77(12):1065–1070, 1962.
93. Theodore, A., Berger, A. G., and Palmer, C. E. A follow-up study of tuberculosis in former student nurses. *J. Chronic Dis.* 3:499–520, 1956.
94. Redmond, C. K., Smith, E. M., Lloyd, J. W., and Rush, H. W. Long term mortality study of steelworkers. III. Follow-up. *J. Occup. Med.* 11:513–521, 1969.
95. Boice, J. D. Follow-up methods to trace women treated for pulmonary tuberculosis. *Am. J. Epidemiol.* 107:127–139, 1978.
96. Paffenbarger, R. S., Wing, A. L., and Hyde, R. T. Physical activity as an index for heart attack risk in college alumni. *Am. J. Epidemiol.* 108:161–175, 1978.
97. Austin, M. A., Berreyesa, S., Elliot, J. L., Wallace, R. B., et al. Methods for determining long-term survival in a population based study. *Am. J. Epidemiol.* 110:747–752, 1979.
98. Talbott, E., et al. Chapter 13, this volume.
99. Talbott, E., Radford, E., Schmeltz, R., Murphy, P., et al. Distribution of thyroid abnormalities in a community exposed to low levels of gamma radiation. *Am. J. Epidemiol.* In press.
100. Pennsylvania Department of Health. Health Surveillance Protocol: Planned Health Activities Related to the Drake Chemical Superfund Site, Lock Haven, Clinton County, Pennsylvania. November 15, 1983.
101. Lyon, J. L., Kauber, M. R., Graft, W., et al. Cancer clustering around point sources of pollution: Assessment by a case-control methodology. *Environ. Res.* 25:29–34, 1981.
102. Blot, W. J., Brown, L. M., Pottern, L. M., Stone, B. J., and Fraumeni, J. F. Lung cancer among long-term steel workers. *Am. J. Epidemiol.* 117:706–716, 1983.
103. Morgenstern, H. Use of ecologic analysis in epidemiologic research. *Am. J. Public Health* 72:1336–1344, 1983.
104. Mason, T. J., McKay, F. W., Hoover, R., Blot, W. J., and Fraumeni, J. F., Jr. Atlas of Cancer Mortality for U.S. Counties: 1950–69. DHEW Pub. No. (NIH) 75-780. Washington DC: U.S. Government Printing Office, 1975.
105. Goodman, L. A. Some alternatives to ecological correlation. *Am. Sociol. Rev.* 64:610–625, 1959.
106. Beval, V., Chilvers, C., and Fraser, P. On the estimation of relative risk from vital statistical data. *J. Epidemiol. Comm. Health* 33:159–162, 1979.

107. Selvin, H. C. Durkheim's "Suicide" and problems of empirical research. *Am. J. Sociol.* 63:607–619, 1958.
108. Langheim, G. I., and Lichtman, A. J. *Ecological Inference.* Beverly Hills, CA: Sage Publications, 1978.
109. Duncan, O. D., Cuzzort, R. P., and Duncan, B. *Statistical Geography. Problems in Analyzing Real Data.* West Point, CT: Greenwood Press, 1961.
110. Valkonen, T. Individual and structural effects in ecological research. In Dugan, M. and Rohbrag, S. (Eds.): *Social Ecology.* Cambridge, MA: MIT Press, 1969.
111. National Technical Information Service. General Catalog of Information Services No. 7. U.S. Dept. of Commerce, 1981.
112. Marsh, G. M., Schaid, D., Sefcik, S., Miller, B., et al. Mortality and Population Data System. Unpublished user manual. University of Pittsburgh, Department of Biostatistics, 1984.
113. Gittelsohn, A. M. On the distribution of underlying causes of death. *Am. J. Public Health* 72:133–140, 1982.
114. Biometry Branch, Division of Cancer Cause and Prevention, National Cancer Institute. SEER Program: Cancer Incidence and Mortality in the U.S. 1973–76. (Young, Jr., J. L., Asive, A. J., and Pollack, E. S., Eds.) DHEW Pub. No. 78-1837, 1978.
115. Greenberg, R., Coltan, T., and Bagne, C. Measurement of cancer incidence in the United States: sources and uses of data. *J. Nat. Cancer Inst.* 68:743–750, 1982.
116. Data Bases in SEEDIS, Lawrence Berkeley Laboratory, Computer Science and Mathematics Department, Berkeley, CA, August, 1982.
117. Merrill, D. W. Overview of Integrated Data Systems. Report LBL 15074. Proceedings of the 1982 Integrated Data Users Workshop, Reston, VA, October, 1982. The American Demographics, 1983.
118. UPGRADE User's Support. A summary of data bases and data collections available through the UPGRADE System. Washington, DC: Sigma Data Computing Corp., April, 1980.
119. UPGRADE User News 3 (Nos. 2–4) 1981 and 4 (No. 1) 1982, Council on Environmental Quality. UPGRADE Project. Washington, DC.
120. McCrea-Curnen, M. G., and Schoenfeld, E. R. Standard information on environmental exposure linked to data on cancer patients, with a brief review of the literature. *Prev. Med.* 12:242–261, 1983.
121. Marsh, G. M. Proportional mortality patterns among chemical plant workers exposed to formaldehyde. *Brit. J. Ind. Med.* 39:313–322, 1982.
122. John, L. R., Marsh, G. M., and Enterline, P. E. Evaluating occupational hazards using only information known to employers. A comparative study. *Brit. J. Ind. Med.* 40:346–352, 1983.
123. DeCoufle, P., Thomas, T. L., and Pickle, L. W. Comparison of the proportionate mortality ratio and standardized mortality ratio risk measures. *Am. J. Epidemiol.* 111:263–269, 1980.
124. Kupper, L. L., McMichael, A. J., Symons, M. J., and Most, B. M.

On the utility of proportional mortality analysis. *J. Chronic Dis.* 31:15–22, 1978.

125. Miettinen, O. S., and Wang, J. D. An alternative to the proportionate mortality ratio. *Am. J. Epidemiol.* 114:144–148, 1981.

126. Freedman, L. S. The use of a Kolmogrov-Smirnov type statistic in testing hypotheses about seasonal variation. *J. Epidemiol. Comm. Health* 33:223–228, 1979.

127. Pike, M. C., and Smith, P. G. Disease clustering: a generalization of Knox's approach to the detection of space-time interactions. *Biometrics* 24:541–556, 1968.

128. Neutra, R. Roles for epidemiology: The impact of environmental chemicals. *Environ. Health Pers.* 48:99–104, 1983.

129. Selikoff, I. J. Commentary, clinical and epidemiological evaluation of health effects in potentially affected populations. *Environ. Health Pers.* 48:105–106, 1983.

130. Cormach, R. M. Numerical taxonomy. *J. Royal. Stat. Soc.* A 134:321–325, 1971.

131. Kleinbaum, D. G., and Kupper, L. L. *Applied Regression Analysis and Other Multivariate Methods.* North Scituate, MA: Duxbury Press, 1978.

132. Jones, J. H., Jones, L., Monson, J. E., Chapman, M., et al. Numerical taxonomy and discrimination analysis applied to non-specific colitis. *Quart. J. Med.* 42:715–732, 1973.

133. Gordon, L. J. Guide to the Role of Local Health Officials in the Local Management of Hazardous Waste. Washington, DC: American Public Health Association, 1982.

134. Sun, M. EPA launches major dioxin attack. *Science* 223 (4631):34, January 6, 1984.

135. SAS Institute, Inc. *SAS User's Guide*, 1979 Edition.

136. Personal Communication. Mr. Steve Caldwell, U.S. EPA Office of Policy and Program Management. Division of Hazardous Site Control, Discovery and Investigation Branch, Washington, DC, October 6, 1983.

137. Barrett, K. W. et al. Uncontrolled hazardous waste site ranking system — a user's manual (MTR-82W111), The MITRE Corp., McLean, Virginia, August, 1982. Also found as National Oil and Hazardous Substances Contingency Plan, Appendix A, *Fed. Reg.*, (47 FR 31219), July 16, 1982.

138. Centers for Disease Control. S.P.A.C.E. for Health: A System for Prevention, Assessment, and Control of Exposures and Health Effects from Hazardous Sites, April, 1983.

139. Kreiss, K., Zack, M. M., Kimbrough, R. D., Needham, L. L., et al. Cross-sectional study of a community with exceptional exposure to DDT. *J. Am. Med. Assoc.* 245(19): 1926–1930, May, 1981.

140. Rothenberg, R. Morbidity study at a chemical dump — New York. *MMWR* 30(24):293–294, June 26, 1981.

141. Halperin, W., Landrigan, P. J., Altman, R., Iaci, A. W., et al. Chemi-

cal fire at toxic waste disposal plant: Epidemiologic study of exposure to smoke and fumes. *J. Med. Soc. of NJ* 78(9):591–594, August, 1981.

142. Parker, G. S., and Rosen, S. L. Cancer Incidence and Environmental Health Hazards, 1969–1978. Woburn, MA: Massachusetts Department of Public Health, January 23, 1981.

143. Personal Communication. Dr. Kent Gray, Centers for Disease Control, Agency for Toxic Substances and Disease Registry, Atlanta, GA, October 6, 1983.

144. National Conference of State Legislatures' Solid and Hazardous Waste Project. Hazardous Waste Management: A Survey of State Legislation, 1982. National Conference of State Legislatures, 1982.

145. Lapham, S. C., and Castle, S. P. (Eds.). A natural perspective on environmental epidemiology. In *Focus on Environmental Epidemiology in New Mexico* (Newsletter), January/February, 1983.

146. Oudbier, A. J., Eyster, J. T., and Lock, J. M. Berlin-Farro Respiratory Study, December, 1981. Michigan Department of Public Health.

147. Conomos, M. G. A Health Census of a Community with Groundwater Contamination, Jackson Township, 1980. New Jersey Department of Health, July, 1983.

148. New Jersey Department of Health. A Health Survey of the Population Living Near Gloucester Environmental Management Services (GEMS) Landfill, April, 1982.

149. New Jersey Department of Health. A Health Survey of Population Living Near the Price Landfill Conducted by the Environmental Health Hazard Evaluation Program, New Jersey Department of Health in Cooperation with the Atlantic County Health Department, July, 1983.

150. Indian, R. W., Campbell, R. J., Holtzhauer, F. J., and Halpin, T. J. Health Survey of a Population in the Proximity of a Chemical Waste Disposal Site, Lorain County. Ohio Department of Health, December, 1980.

151. Paigen, B., and Goldman, L. R. Chapter 9, this volume.

152. Heath, C. W. Chapter 11, this volume.

153. Picciano, D. Pilot cytogenetic study of the residents living near Love Canal, a hazardous waste site. *Mammalian Chromosome Newsletter* 21:86–93, 1980.

154. Kolata, G. B. Love Canal: False alarm caused by botched study. *Science* 208:1239–1242, 1980.

155. Clark, C. S., Meyer, C. R., Gartside, P. S., et al. An environmental health survey of drinking water contamination by leachate from a pesticide waste dump in Hardeman County, Tennessee. *Arch. Environ. Health* 37(1):9–18, January/February, 1982.

156. Meyer, C. R. Liver dysfunction in residents exposed to leachate from a toxic waste dump. *Environ. Health Pers.* 48:9–13, 1983.

CHAPTER 2

Risk Perception in Waste Disposal:
A Historical Review

Joel A. Tarr

INTRODUCTION

This chapter is concerned with changes in attitudes toward wastes and methods of waste disposal over time. For our purposes, waste is defined as something "left over or superfluous." This definition, however, should not foreclose the possibility of reuse or recycling. Our attitudes toward wastes and our methods of disposing of them or reusing them, are affected by a host of cultural, economic, and technical factors. Normally we object to wastes for health or aesthetic reasons, but perspectives concerning what is or is not objectionable have shifted markedly. What is believed to be a health danger in one period of time or in one culture may be viewed differently in another time or by another culture: what one society sees as a nuisance to be eliminated, another may perceive with indifference.[1] What we are primarily dealing with, therefore, are attitudes toward risk in regard to wastes from either a health or socially related perspective.

Normally we conceive of primarily two kinds of waste streams. One waste stream is the product of everyday living activities: food wastes or garbage; refuse, such as paper or ashes (nonorganic materials); and human body

J. B. Andelman and D. W. Underhill (Editors), *Health Effects from Hazardous Waste Sites*
© 1987 Lewis Publishers, Inc., Chelsea, Michigan — Printed in U.S.A.

wastes (feces and urine). The second waste stream derives largely from the various productive activities in which a society engages, including agriculture, raw materials processing, and manufacturing. The great bulk of waste materials is actually generated by the second set of processes although individuals tend to be most aware of the first waste stream. These waste products may be in either liquid, solid, or gaseous form and may be disposed of in any or all of the environmental media: air, land, or water.

For the purposes of this chapter, waste disposal will be viewed in terms of its perceived health effects. Here, there are clear stages of development and change related to new hypotheses of disease etiology and available measurement instruments. Over time, alterations in these areas as well as in the value systems of users, resulted in important policy changes and new practices in regard to waste disposal.

THE AGE OF THE MIASMAS

For a considerable part of the nineteenth century, the two dominant theories of disease etiology were the contagionist and anticontagionist theories. Briefly, contagionists argued that epidemics were caused by specific contagia that were transmitted from individual to individual, usually originating from foreign sources. Hence, the proper policy to pursue once diseases such as cholera, typhoid, or yellow fever were identified was to declare a quarantine and close off the nation, region, city, or state to suspected carriers. Contagionism probably reached its peak in the 1850s with the National Quarantine Conventions. From the 1850s through the 1880s, an anticontagionist hypothesis, known as the "filth," "pythogenic," or "miasmic" theory, was most widely accepted. Essentially, this hypothesis held that infectious or "zymotic" disease evolved de nova from putrefying organic matter and resulted in epidemics of cholera, yellow fever, or typhoid. The logical policy outcome of this hypothesis was to remove organic matter from the cities before it decomposed.[2]

The filth theory of disease was embodied in the nineteenth century movement known as the Sanitary Movement. This movement began in Great Britain with the work of Sir Edwin Chadwick and his followers to promote a healthful urban environment by cleansing the cities. It can be viewed as a social movement led by elites and professionals who aimed to change people's ideas about personal hygiene. Chadwick's vision included an expanded role for government in areas related to health and sanitation and the construction of public works to achieve a healthful city. The motivation of the sanitationists was more than humanitarian — improved sanitary conditions were necessary to improve workers' health and protect the health of the community.[3]

Chadwick's ideas greatly influenced the pioneer group of American sanitarians and public health reformers. The institutional and organizational embodiments of the Sanitary Movement in the United States were the American Public Health Association (1871), the short-lived National Board of Health (1879–1883), and the multitude of local and state boards of health that appeared in the late nineteenth century.[2] Municipalities and states passed laws regulating a range of activities related to health and sanitation, such as cesspool and privy construction and emptying, street cleaning, garbage collection, sewerage development, and water supply. Spokesmen for the Sanitary Movement believed technology essential to solve the urban waste removal problem and pushed for municipal adoption of capital-intensive sewerage systems to speed human waste products from the home and the city.[2,4]

Pythogenic or miasmic theory held that all organic waste matter was suspect, and sanitarians placed equal emphasis on cleaning the streets of horse manure and controlling slaughter houses and burial grounds, as on regulating privies. This attitude also resulted in a concern with organic trade wastes from tanneries, textile works, paper mills, and cigar companies. The first state legislation to control stream pollution was an 1878 Massachusetts law giving the State Board of Health the power to control the river pollution caused by organic manufacturing wastes.[5-6] In addition, by the 1880s many municipalities had passed statutes restricting so-called noxious manufacturers, such as slaughterhouses and gas works, to the fringes of the cities, as well as legislation regulating the construction and cleaning of cesspools and privy vaults and garbage disposal. By 1880, most cities with above 30,000 population had a board of health, a health commission, or a health officer, about half of whom had direct control over the collection and disposal of refuse.[7-8]

Of 99 cities with over 30,000 population surveyed in the 1880 U.S. census, the most common method utilized to dispose of garbage, street sweepings, and ashes was dumping it on the ground. Usually dumps were on the urban fringe, but vacant lots within the city boundaries were occasionally used. Garbage and refuse were also often placed in landfills such as Boston's Back Bay or along Manhattan's shoreline. A number of cities disposed of their wastes as animal feed or fertilizer, although this required some presorting. A relatively small number of cities (but often large in size) dumped their refuse into adjacent waterways. New York City, for instance, disposed of most of its refuse in the ocean, while Chicago often used Lake Michigan.[8] Most cities had no specific regulations regarding the disposal of manufacturing waste but dealt with them under nuisance provisions of the law. Few cities in 1880 had sanitary sewer systems, and human body wastes were deposited in privy vaults and cesspools that required periodic emptying. The same means of disposal were used for human body wastes as for garbage and refuse. It is worth noting that the privy wastes of over 100 cities were used as fertilizer in 1880.[9]

THE BACTERIAL REVOLUTION AND THE REVERSAL OF ATTITUDES CONCERNING HEALTH AND WASTE DISPOSAL

Leading sanitarians believed that the city's health problems would be solved by installing water carriage technology that would remove human wastes from the household to a remote place for disposal. In the large majority of locations, this place was a nearby waterway. Essentially this involved the shifting of wastes from a land disposal sink to a water sink. Underlying the use of waterways for ultimate disposal was the concept that running water purified itself—a hypothesis often confirmed by existing means of chemical analysis. While this theory is not entirely incorrect, it is operative only under certain conditions of flow and mixing. Invariably, the result of dumping raw sewage into streams from which downstream cities drew their water supply was a large increase in morbidity and mortality rates from infectious disease such as typhoid fever for the downstream communities.

In the 1890s, bacterial researchers, following the seminal work of Pasteur and Koch in establishing the germ theory, identified the processes involved in waterborne disease. The work of William T. Sedgwick and other bacterial researchers at the Massachusetts Board of Health's Lawrence Experiment Station was especially critical in clarifying the etiology of typhoid fever and confirming its relationship to sewage-polluted waterways. The irony was clear: cities had adopted water carriage technology because of an expectation of local health benefits resulting from more rapid and complete collection and removal of wastes, but disposal practices produced serious externalities for downstream or neighboring uses.[10]

The challenge of waterborne infectious disease to public health was met by the development of two further technologies—water filtration and chlorination in the pre-World War I period. By 1940 almost all urbanites were drinking treated water, and morbidity and mortality from waterborne disease had ceased to be a serious public health problem. Sewage treatment, however, which was intended primarily to deal with sewage-created nuisances in waterways, lagged far behind. One result of focusing on the treatment of water at the intake point rather than the treatment of sewage before it entered the receiving body of water was the degradation of waterways for recreational purposes.[10]

Compared to sanitary wastes from human populations, industrial wastes were relatively neglected during this period. Because they normally did not contain disease germs, public health authorities argued that "their relation to sanitation is remote."[11] Thus the shift from miasmic theory to bacterial theory reduced the concern with the health effects of industrial and especially organic industrial wastes. In the 1920s and 1930s, sanitary engineers identified the following as the main problems caused by industrial wastes: interference with water filtration systems, oxygen consumption characteristics that made streams odorous and unusable for potable purposes, the creation of taste problems in drinking water (phenols especially), and devastating effects on fish life (taste, BOD, and toxic effects, especially by gashouse wastes).[12] "Few wastes," however, noted one authority in 1938, "are present in most streams in sufficient quantities to become poisonous."[13] In fact, in regions such as western Pennsylvania with a high incidence of mine-acid drainage and steel mill pickling-liquor discharge, engineers argued that these wastes had a germicidal effect on sewage pollution and therefore should not be excluded from streams.[14]

The replacement of the filth theory with the germ theory caused public health authorities to reduce their interest in solid waste collection and disposal. Leaders of the so-called New Public Health argued that public health officials should focus on the control of the diseased or carrier individual rather than environmental sanitation. This reorientation caused many municipalities to transfer control of refuse collection and disposal from health departments to sanitation or public works departments. Solid waste disposal was now viewed as an engineering problem involving nuisance and cost considerations rather than an issue with public health implications.[8]

Questions concerning the means of disposal largely revolved around the cost-effectiveness of different forms of technologies. Incineration and reduction were widely used, although land dumping and hog feeding also remained popular until after World War II. Ocean and waterway dumping, however, were increasingly abandoned because of nuisance, cost, and legal considerations. In 1934, as a result of a United States Supreme Court decision, dumping of municipal wastes at sea was ended. All of the above mentioned techniques, however, were to be seriously challenged after the war.

POST-WORLD WAR II WASTE DISPOSAL TRENDS

In the postwar decades, through the landmark environmental legislation of the 1970s, the prime focus was on water and air pollution, with an increasing concern for land pollution after 1965, but especially since the late 1970s. In regard to water pollution control, there has been steadily increasing federal authority, dating from congressional passage of the Federal Water Pollution Control Act in 1948 through the enactment of the 1972 amendments, more popularly known as the Clean Water Act (PL 92–500). The concepts and goals embodied in this legislation — uniform water quality standards, such as zero effluent and fishable, swimmable waters — had actually appeared in earlier state enactments such as the Pennsylvania Clean Streams Act of 1905, but never with enforceable provisions. The 1972 legislation, in contrast, embodied the strongest sanctions up to that time.

The strong sanctions enacted in regard to water pollution, as well as those applied to air pollution by the Clean Air Act, had an unpredicted effect.

Because surface waters and the air were no longer acceptable sinks for the disposal of wastes, industries turned increasingly to the land. The 1979 report of the Council on Environmental Quality noted, for instance, that "the increasing tempo of the cleanup of lakes and streams is literally driving pollution underground."[15] That this would occur could have been foreseen from past experience. In the 1940s, for instance, when the Pennsylvania Sanitary Water Board began enforcing the state's Clean Streams Act, a number of small industrial plants turned to the use of earthen lagoons on plant property as a means of avoiding controls. These lagoons, many of which were poorly constructed and unlined, ultimately threatened groundwater supplies, posed nuisances, and even created air pollution problems.[16]

Deep-well injection is another method of land-based industrial waste disposal that expanded because of regulations restricting disposal in surface waters. The expanding chemical and petroleum industries, facing serious disposal problems, developed this technique in the 1930s, and its use grew in the postwar years. In 1960 there were about 30 deep wells throughout the country, but in that decade, because of enforcement of water and air pollution statues by the various levels of government, the number increased to 110. The concept behind a properly designed deep well was that it would take the "effluent out of the human environment and bury it forever," but many firms located the wells in strata where the wastes they contained posed a threat to underground aquifers and drinking water supplies.[17]

Over the years, industries have actually punched or dug thousands of holes in the ground, usually on their own property, to dispose of wastes. According to a 1979 article:

> Landfilling has long been the most common method for disposal of hazardous wastes because it has been inexpensive. . . . The costs were low because the technology was simple. Typically, a hole was dug in clay at a selected site, unconsolidated sludge and drums of chemicals placed in it, and the hole was filled and covered with clay to keep out rain and other water.[18]

A review of the professional literature in the area of waste disposal confirms that the above quotation accurately describes the procedures followed, although information in the published literature is very limited. As a chemical plant manager noted in 1967, the practice in regard to hazardous materials was to vent them to the atmosphere, pour them down the drain, or, "if it is very obnoxious or messy, drum it up, set it out in the back lot, and forget about it."[19] Another article of the same period noted that several chemical companies had discovered forgotten chemical dumps on their own property when expanding plant facilities.[20] As noted sanitary engineer Abel Wolman has observed, the handling of hazardous wastes previous to 1976 was "haphazard, desultory and . . . certainly not carefully reviewed, designed or operated." Such sloppy "housekeeping practices," he says, were a result of a desire to dispose of wastes in the simplest and cheapest manner.[21] Careless "housekeeping" practices had increasingly serious implications for environmental conditions, as the volume and variety of hazardous wastes produced by industry greatly expanded in the postwar period. Also important was population growth and physical expansion of metropolitan areas, causing the encroachment of residences and commercial activities on industrial production and disposal sites.

In the postwar period, municipalities as well as industries began using the land more intensely for waste disposal. The technique they utilized was the sanitary landfill, a method of garbage and refuse disposal originally developed in Great Britain around World War I called "tipping." A few American cities, such as Fresno, San Francisco, Seattle, and New York, developed landfills in the 1930s, but it was not until after the war that the technique became widespread. In the 1930s, municipalities rejected solid waste disposal methods, e.g., incineration, hog farms, and reduction, as well as open dumping, for economic, health, and nuisance reasons, and replaced them with sanitary landfills.[8]

Sanitary landfill was a method of solid waste disposal that involved the filling of depressions or trenches in the ground with refuse, and using a bulldozer or a bull clam shovel to dig the trench and compact the fill. Each day's deposit was sealed in an individual refuse cell. When the fill reached the desired level, it was covered with earth and again compacted. The land created by a completed fill was often used for recreational or even building purposes.[22]

Most important, the use of sanitary landfills appeared to eliminate the nuisances produced by other methods of disposal and apparently produced no public health hazards. The standards required for the proper use of the sanitary landfill technique were set forth in 1938 by a committee of public health specialists and sanitary engineers headed by Dr. Thomas Parran, Surgeon General of the United States. The New York Supreme Court had established the committee as a means to adjudicate a challenge by a group of Queens, New York citizens to the establishment of a landfill on Jamaica Bay. After several weeks of investigation, the Parran committee found the sanitary landfill method to be free of dangers to public health or safety. Sanitary landfill, noted the committee report, was a significant health improvement over the open dump, eliminated undesirable marsh and swamp land (today called "wetlands") that harbored rats and mosquitoes, and provided a benefit in terms of filled-in ground. The commission considered several possible landfill hazards, such as fires and low weight-bearing value, but maintained that they could be controlled by proper precautions. No mention was made, however, of other possible dangers from leachate runoff or groundwater pollution, or the possibility of long-term health hazards. Most of the commission's discussion of risk was in terms of nuisance rather than health danger. The report enumerated fifteen rules, mostly operational, for the safe conduct of sanitary landfill operations.[23–25]

The Parran Report, combined with a favorable experience by the army with sanitary landfills at its camps in the United States during World War II, gave the technique wide appeal in the postwar period. Public works officials and public health professionals strongly endorsed it as a method of waste disposal and especially as a replacement for the open dump. The advantages most commonly cited were those listed by the Parran Commission: the elimination of the nuisances and health hazards associated with open dumps, the filling in of marshes and swamps with a consequent reduction of rats and mosquitoes, and the creation of land for buildings, parks and recreational areas. Isolated articles did appear in the 1950s that noted hazards at operating landfills, such as methane fires, groundwater pollution, and low weight-bearing capabilities that restricted building, but most journals focused on the benefits of the technique and not the risks.[26]

In 1961 the Sanitary Engineering Division of the American Society of Civil Engineers (ASCE) published a survey of sanitary landfill practice that examined 250 sites. Completed landfills were most commonly used for recreational and industrial purposes, although some fills were used for homesites and schools. Of the fills surveyed, 12% were less than 250 feet from the nearest dwelling. The article noted that while groundwater pollution from landfills was a "critical item," in general site planning it was given minimal concern, with 79% of the sample within 20 feet of groundwater and 27% at or near groundwater. Only 9.3% of the sites reported that operators had made test ground boring prior to fill operations, and only 14% had specially engineered drainage devices. The survey reported that, in spite of the purported safety of landfills, citizens often opposed having them located near their residences.[27]

The 1961 survey also noted that over 70% of the landfills examined operated under some sort of municipal or county regulations. During the 1950s, as landfills became more common, cities issued sanitary landfill regulations; states such as California and Illinois suggested operational guidelines; and professional groups, especially the American Public Works Association, the Sanitary Engineering Division of the American Society of Civil Engineers, and the U.S. Public Health Service, conducted investigations on standards to avoid undue risk. By the time of the ASCE survey, professional groups involved in solid waste questions agreed that while sanitary landfills reduced disposal costs and were superior to the open dump, they still presented dangers in regard to leachate seepage; groundwater pollution; poor load bearing; methane; and nuisances such as rats, vermin, and blowing paper. A lack of research, however, restricted the availability of technical information on these hazards that could be used to refine practice.[28]

In 1963, in an attempt to generate interest and research in the solid waste area, the U.S. Public Health Service and the American Public Works Association sponsored the First National Conference on Solid Waste Research. In his keynote address, Professor J. E. McKee of the California Institute of Technology offered four explanations for the lack of research in solid waste disposal. First, McKee noted, neither cities, regulatory agencies, nor the public demanded such information. Second, solid waste had produced no public health crises equivalent to those in air and water pollution that could have generated such a demand. Third, federal and state government was minimally involved in the area. And fourth, most officials and engineers concerned with solid waste disposal considered it an economic and political rather than a scientific or engineering problem.[29] All of these factors applied to the sanitary landfill technique as well as to solid wastes in general, but landfills did not hold an especially prominent place at the conference. In addition, neither the papers on sanitary landfill, nor the discussions following their delivery emphasized possible hazards. Concern for the land as a sink for pollutants did not yet possess the urgency that was beginning to characterize disposal in air and water in the 1960s.

THE GROWTH OF AWARENESS REGARDING THE HAZARDS OF WASTE DISPOSAL IN LAND

Conferences, such as the one sponsored in 1963 by the Public Health Service and the American Public Works Association, highlighted the deficiencies in solid waste research and suggested the need for legislation in the area. More critical for federal involvement, however, was the growing expense of solid waste collection and disposal. Powerful urban politicians, including Chicago's Mayor Richard Daley, pushed for federal action to lighten the burden on cities. In 1965, after President Lyndon B. Johnson had spoken of the need for "better solutions to the disposal of solid waste" and called for legislation, the U. S. Congress passed the Solid Waste Disposal Act. This act created the Office of Solid Wastes and provided the federal government with a more formal role in regard to municipal wastes.[8]

The Solid Waste Disposal Act provided funds for research, investigation, and demonstration in the solid waste domain and for technical and financial assistance to state and local governments and interstate agencies in "the planning, development and conduct" of disposal programs. The program's most important impacts were to stimulate research and to inspire state government activity. In 1965, for instance, there was no state level solid waste agency in the country, but by 1970, 44 states had developed programs.[8] During the 1970s, however, the focus of federal legislation moved from research into conventional methods of solid waste disposal toward the reuse and recycling of resources. This was reflected in the passage of the Resource Conservation and Recovery Act of 1976 in the form of amendments to the 1965 legislation.

Section 212 of the 1970 Solid Waste Act required that the U.S. Environmental Protection Agency undertake a comprehensive investigation of the storage and disposal of hazardous wastes. This led to a report to Congress in 1974 on the disposal of hazardous wastes and eventually, in 1976, to the passage of the Resource Conservation and Recovery Act. The act attempted to fill the regulatory gaps concerning the disposal of hazardous wastes left by state programs. In addition, early in 1980, acting under the requirements of RCRA, the EPA announced new regulations implementing cradle-to-grave controls for handling hazardous wastes.[15,30] The use of the land as a sink was now to be finally curtailed.

The various acts passed after 1965 and investigations conducted under their authority caused a convergence of the different streams of research concerning municipal wastes on the one hand and industrial hazardous wastes on the other. The point of convergence was landfill type operations, with special concern over site construction and hazardous groundwater pollution. In 1979 the Environmental Protection Agency estimated there were about 75,000 active industrial landfills while a 1978 *Waste Age* survey identified more than 14,000 active municipal landfills.[31] The number of abandoned landfills is unknown.[30]

The most serious potential problem involving both municipal and industrial landfills is the threat to groundwater quality. Relatively little attention had been paid to this question before World War II, although there were a few isolated warnings of potential dangers. In the postwar years, several state departments of health (California was especially active) issued warnings about potential chemical pollution of groundwater from sanitary fills. Studies concerning the effects of industrial effluents such as metal plating wastes, phenols, oil field brines and chemical products on groundwater began to appear in the literature. When landfill research accelerated during the 1960s, so did awareness about the hazards of possible groundwater pollution.[32] By 1970 many states had regulations requiring field investigations of groundwater location in the siting of new municipal and industrial landfills. Where problems did exist, however, they usually centered around older sites that had been developed without adequate investigation of the risk of possible groundwater contamination.

There are several different reasons why the potential for groundwater contamination from landfills had been neglected before the 1970s. One is the lack of research in the area of solid waste disposal in general and landfills in particular, as well as a basic ignorance about underground processes. Writing in 1972, one author noted that "before 1965 very few people were aware of the fact that water passing through refuse in a landfill could become highly contaminated . . . few cases were noted where leachate had caused harm to someone."[33] Research and publication regarding the disposal of industrial wastes was also restricted "owing to the competitive nature of private enterprises, and their reluctance at times to divulge operating problems and techniques."[39] In addition, a lack of analytical instrumentation making possible the tracing of landfill contaminants or the detection of extremely low levels of potentially hazardous substances provided a restraint on knowledge. Before federal legislation, there was no incentive system to spur research in either analytical chemistry in regard to groundwater processes or groundwater-leachate-soil exchanges.[35]

Equally important was the absence of a clear hazard or crisis involving groundwater pollution from waste disposal. As one sanitary engineer noted in 1968, a "major obstacle to the solution of solid waste problems is the lack of an awareness on the part of governmental decisionmakers that the problem even exists. This lack of awareness exists at all levels."[36] Up to 1970, few incidents involving the pollution of groundwater drinking supplies by wastes had been reported, and municipalities and state governments often ignored the problem. Rather than spend scarce funds on expensive testing and monitoring, governments put their dollars in areas where need appeared more immediate.

CONCLUSIONS

The body of federal and state environmental law that has appeared in the last two decades reflects a new set of values. These values include a regard for the quality of the natural environment from an amenity and aesthetic perspective and a concern for health in relationship to the environment.[37] These new values also reflect the findings of the emerging discipline of environmental health, with its focus on the chronic, degenerative diseases.[38] In this regard, we have returned to the original thrust of the nineteenth century Sanitary Movement toward cleansing the urban environment in order to ensure freedom from epidemics.

This review of the history of waste disposal has suggested that environmental degradation and the ignoring of ill effects were not always a result of a willful act on the part of the waste generator. An environmental hypothesis such as "running water purifies itself" provided sanction to cities to dispose of their sewage in nearby streams, while chemical analysis seemingly gave a "scientific" stamp of approval. Ringlemann charts supplied a method of grading smoke pollution but did not identify other insidious but invisible air pollutants. Sanitary and industrial landfills produced leachates that contaminated groundwater used for drinking water supplies, but limited monitoring capabilities hindered detection. Research in these areas often only developed after the occurrence of crisis situations and as a result of specific public policies and not before. But even after research had pinpointed the mechanisms by which capital technologies such as sewers or landfills produced negative effects, it often proved difficult to persuade the operators of these technologies, be they private or public, to cease using the polluting technology or to stop building new systems having the same effects. As a result, we must simultaneously deal with the results of careless past waste disposal practices as well as the difficult waste disposal problems of our own time.

REFERENCES

1. Douglas, Mary. Purity and Danger. Basic Books, New York, 1966.
2. Rosen, George (1958). A History of Public Health. MD Publications, New York, 1958.
3. Sheppard, Francis. London 1808–1870: The Infernal Wen. University of California Press, Berkeley, 1971.
4. Tarr, Joel A. and McMichael, Francis C. The Evolution of Wastewater Technology and the Development of State Regulation: A Retrospective Assessment. In Joel A. Tarr and F. C. McMichael, Eds. Retrospective Technology Assessment, pp. 165–190, 1977.
5. Massachusetts State Board of Health. An Inquiry into the Causes of Typhoid Fever as it Occurs in Massachusetts. In Second Annual Report of the State Board of Health of Massachusetts. Wright and Potter, Boston, pp. 110–179, 1872.
6. Rosenkrantz, Barbara. Public Health and the State; Changing Views in Massachusetts, 1842–1936. Harvard University Press, Cambridge, 1972.
7. Chapin, Charles V. History of State and Municipal Control of Disease. In Mazyck P. Ravenel, Ed. A Half Century of Public Health. American Public Health Association, New York, pp. 133–160, 1921.
8. Melosi, Martin V. Garbage in the Cities: Refuse Reform, and the Environment. Texas A&M University Press, College Station, Texas, 1981.
9. Tarr, J. A. From City to Farm: Urban Wastes and the American Farmer. *Agricultural History*, 49:598–612, 1975.
10. Tarr, Joel A., McCurley, James III, and Yosie, F. Terry. The Development and Impact of Urban Wastewater Technology: Changing Concepts of Water Quality Control, 1850–1930. In Martin V. Melosi, Ed. Pollution and Reform in American Cities, 1870–1930, pp. 59–82, 1980.
11. Leighton, M.O. Industrial Wastes and Their Sanitary Significance. Reports and Papers of the American Public Health Association 31:29–41, 1905.
12. Besselievre, E. B. The Disposal of Industrial Chemical Waste. *Chemical Age* 25:516–518, 1931.
13. Warrick, L. F. The Prevalence of the Industrial Waste Problem. In Langdon Pearse, Ed. Modern Sewage Disposal, 340-372, 1938.
14. Pennsylvania State Department of Health. The Germicidal Effect of Water from Coal Mines and Tannery Wheels Upon Bacillus Typhosus, Bacillus Coli and Bacillus Anthracis. Pennsylvania Health Bulletin 5:1–10, 1909.
15. Council on Environmental Quality. Environmental Quality, 1979. U.S. Government Printing Office, Washington, DC, 1979.
16. Lazarchik, Donald A. Pennsylvania's Pollution Incident Prevention

Program. Proceedings of the 25th Annual Purdue Industrial Waste Conference 25:528–533, 1970.

17. Walker, William R. and Stewart, Ronald C. Deep-Well Disposal of Wastes. Journal of the Sanitary Engineering Division, Proceedings of the American Society of Civil Engineers 94:945–1068, 1968.

18. Maugh, Thomas H. Burial is Last Resort for Hazardous Wastes. *Science*, 204:1294–1297, 1979.

19. W. M. Deviny. Pilot Plant Procedures; Disposal of Hazardous Chemicals. *Chem. Eng. Prog.* 63:56–57, 1967.

20. Chemical World. Cash to Counter Pollution. *Chem. World* 114, October 29, 1969.

21. Hollander, Walter (Comp.). Abel Wolman: His Life and Philosophy. Universal Printing & Publishing Co., Chapel Hill, NC, 2 vols., 1981.

22. American Public Works Association (APWA). Municipal Refuse Disposal. American Public Works Association, Chicago, second ed., 1966.

23. Engineering News Record. Health Experts Endorse Landfills and Recommend Best Practice. *Eng. News Rec.* pp. 54–55, Mar. 28, 1940.

24. Rice, John L. and Pincus, Sol. Health Aspects of Land-Fills. *Am. J. Public Health* 30:1391–1398, 1940.

25. Eliasson, Rolf and Lizee, Albert J. Sanitary Landfills in New York City. *Civ. Eng.* 12:483–486, 1942.

26. Tarr, Joel A. The Search for the Ultimate Sink: Urban Air, Land, and Water Pollution in Historical Perspective. Records of the Columbia Historical Society 51:16–26, 1984.

27. Committee on Sanitary Engineering Research, American Society of Civil Engineers. A Survey of Sanitary Landfill Practices. Journal of the Sanitary Engineering Division. Proceedings of the ASCE 87:65–83, 1961.

28. American Public Works Association (APWA). Solid Wastes Research Needs. American Public Works Association [APWA], Chicago, 1962.

29. American Public Works Association (APWA). Proceedings: National Conference on Solid Waste Research. American Public Works Association, Chicago, 1963.

30. Council on Environmental Quality. Environmental Quality, 1980. U.S. Government Printing Office, Washington, DC, 1980.

31. Greenberg, M. R. and Anderson, R. F. Hazardous Waste Sites: The Credibility Gap. Center for Urban Policy Research, New Brunswick, NJ, 1984.

32. Todd, David Keith and McNulty, Daniel E. Orren. Polluted Groundwater: A Review of the Significant Literature, Huntington, NY: Water Information Center, 1976.

33. Boyle, W. C. and Ham, R. K. Treatability of Leachate from Sanitary Landfills. Proceedings of the 27th Annual Purdue Conference on Industrial Wastes 27:687–691, 1972.

34. Barnes, George E. Industrial Waste Disposal. *Mech. Eng.* 69:465–470, 1947.

35. Brooks, Harvey. Science Indicators and Science Priorities. *Sci., Technol. Human Val.* 38:14–31, 1982.
36. Ludwig, Harvey F. and Black, Ralph J. Report on the Solid Waste Problem. Journal of the Sanitary Engineering Division, Proceedings of the American Society of Civil Engineers 94:355–370, 1968.
37. Hays, Samuel P. From Conservation to Environment: Environmental Politics in the United States Since World War II. *Environ. Rev.* 6:14–41, 1982.
38. Lippmann, M. and Schlesinger, R. B. *Chemical Contamination in the Human Environment.* Oxford University Press, New York, 1979.

SECTION II

Assessment of Exposure

CHAPTER 3

Recent Trends in Environmental Monitoring Near Hazardous Waste Sites

Glenn E. Schweitzer and Shelly J. Williamson

INCREASING ATTENTION TO MONITORING APPROACHES

The enactment of the Resource Conservation and Recovery Act (RCRA) and the Comprehensive Environmental Response, Compensation and Liability Act (CERCLA) has greatly increased the attention of the scientific community to the importance of monitoring near hazardous waste sites. These laws and subsequent regulations require certain types of groundwater monitoring, although the details of most monitoring activities are defined on a site-by-site basis.

A wide variety of methods which have not been fully evaluated are being used in operational monitoring programs due to the urgency of hazardous waste problems. These methods are being documented and refined on a continuing basis, and indeed upgrading monitoring methods will be a high priority for the foreseeable future.

Concomitant with this effort to improve monitoring methods has been a major expansion of quality assurance activities, particularly with regard to chemical analyses in the laboratory. The first priority has been to assure that data of *known* quality are generated. This requirement involves

J. B. Andelman and D. W. Underhill (Editors), *Health Effects from Hazardous Waste Sites*
© 1987 Lewis Publishers, Inc., Chelsea, Michigan – Printed in U.S.A.

detailed documentation of the protocols that are used, the development of data on achievable precision and accuracy for each method and for each set of analyses, and the determination of minimum detection limits. Concerns over sample representativeness, integrity, and contamination are high, and a host of procedures for upgrading field sampling procedures and laboratory performance are being put in place.

Most aspects of the integrity of monitoring data have been discussed at previous conferences and in the technical literature. (See, for example, *Management of Uncontrolled Hazardous Waste Sites*, Proceedings of National Conference, Nov. 29-Dec. 1, 1982, Washington, DC, U.S. EPA et al.) Therefore, this presentation is directed to other less well-publicized developments in the monitoring field which have been stimulated, at least in part, by concerns over chemicals escaping into the environment from hazardous waste sites that could have an impact on human health.

Each waste site is different. Different quantities and combinations of chemicals are present in different types of facilities. Each site is uniquely configured in a unique geographical setting. Still there are a few common concerns which are the themes of this presentation: determining levels of contamination near the sites, objectivity in the acquisition and interpretation of monitoring data, relating monitoring data to the population at risk, and linking environmental monitoring measurements to exposure levels that are potentially harmful to the public. The presentation concludes with a few observations as to how monitoring data can be applied in supporting health research studies.

SUBSURFACE MONITORING

Subsurface monitoring techniques have received unprecedented attention as the result of the growing concern over groundwater contamination near hazardous waste sites. A variety of techniques developed primarily for geological and hydrological investigations and used by the mining and petroleum industries are being adapted for use in environmental assessments.

During the past decade, geological coring, groundwater monitoring wells, and laboratory analyses of soil and water samples have generally characterized the approach for providing information concerning subsurface contamination near hazardous waste sites. While direct sampling will continue to provide much of the definitive information about contamination patterns, a number of complementary technologies are being refined which should aid in detecting and characterizing subsurface pollution problems and in determining the location and frequency of sampling activities. Attention is also being directed to "early warning" detection of pollutant movement at the edge of waste sites and within the unsaturated zone above the water table, thus facilitating corrective action before groundwater resources are adversely impacted.

Three particularly promising developmental areas for expediting and improving subsurface investigations are geophysical techniques, sampling for organic chemicals in the unsaturated zone, and improved approaches to groundwater sampling.

Surface Geophysics and Downhole Sensing

Table 1 shows the potential roles of six surface geophysical methods in hazardous waste site assessments. The methods are complementary, and no method can be used for all applications. In general, metal detectors and magnetometers are most useful in locating buried wastes; ground penetrating radar is the technique of choice for defining the boundary of buried trenches; electromagnetic induction and resistivity are the most useful in defining plumes of conductive contaminants in groundwater; and resistivity and seismic techniques are most useful in determining geologic stratigraphy.

Borehole monitoring systems are useful in obtaining insights as to the vertical lithology, porosity, and structure of the subsurface geology and to the delineation of the saturated and unsaturated zones. Also, they can help characterize subsurface contaminant plumes and determine groundwater flow velocity and direction. Among the most promising downhole techniques are electromagnetic induction, natural gamma logging, fluid conductivity and temperature measurements, and measurements of groundwater flow. All of these techniques now need to be better calibrated in actual field conditions. Meanwhile, they are being deployed on a limited basis to provide information for improved targeting of groundwater sampling activity.

Monitoring in the Unsaturated Zone

Soil coring techniques using hand- or power-driven augers are well established direct monitoring methods. The technical questions in field operations usually relate to the depth of sampling, compositing of individual cores, prevention of cross contamination among cores, and preservation of cores, particularly if analysis for volatile organic chemicals is planned. However, protocols for laboratory analyses of soil samples are only now being adequately evaluated, and technical problems remain for some types of analyses.

Suction cup lysimeters have been used for many years to detect solutes beneath irrigated fields and pollutants beneath sanitary landfills. Usually the concern has been with metals and inorganic salts that may be leaching through the soil at depths of a few feet below the surface. Improved evaluation and documentation are now needed of lysimeter performance in monitoring organic as well as inorganic compounds using different types of cups, such as Teflon, in different soil types and soil conditions.

Table 1. Potential Applications of Surface Geophysics to Hazardous Waste Site Assessments

	Mapping Hydrogeologic Features	Mapping Contaminant Plumes	Locating Trench Boundaries (w/o metal targets)	Locating Trench Boundaries (w/metal targets)	Locating Metal Objects
Ground Penetrating Radar: Identifies interfaces of materials having different electrical properties	good	fair	good	good	fair
Electromagnetic Induction: Measures conductivity of subsurface features including soil, rocks, groundwater, and other porous fluids	good	good	good	good	fair
Electrical Resistivity: Measures resistivity of subsurface features	good	good	fair	fair	
Seismic Refraction: Responds to density, thickness, and depth of materials with different acoustic transmission properties	good		fair	fair	
Metal Detector: Responds to changes in electrical conductivity due to metallic objects				fair	good
Magnetometer: Responds to distortions in earth's magnetic field				fair	good

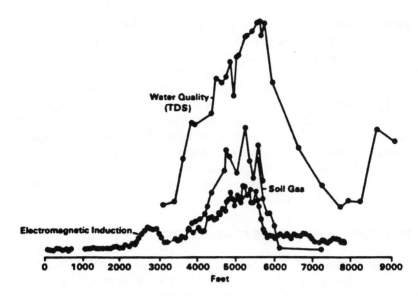

Figure 1. Subsurface sampling for organic vapor.

Evaluations of available sampling techniques for measuring gaseous emissions in the soil column have recently been initiated. Isolation flux chambers for capturing emissions at the surface or at varying depths are designed to allow estimation of emission rates as well as measurement of concentrations that are present. Subsurface probes containing Draeger® tubes can measure concentrations, or portable gas chromatographs can be used to analyze gas samples collected from subsurface probes to establish horizontal profiles of concentrations at selected depths. These techniques are still largely in the experimental stage.

Figure 1 presents an example of the results of a limited subsurface sampling effort to obtain a profile of contaminant concentrations at a depth of 10 to 15 feet, just above a shallow aquifer contaminated with an organic solvent. The soil gas profile correlates very well with data on the extent of the contaminant plume obtained from groundwater wells in the area. The profile also exhibits good correlation with geophysical measurements of the presence of subsurface contamination from metallic ions, thus suggesting that the organic and inorganic contaminants are moving as a single plume.

Groundwater Sampling

Given the large number of organic contaminants and the low concentration levels that are becoming of concern in groundwater, increased attention is being directed to the possibilities of inadvertent contamination during sampling activities. The selection of monitoring well construction materials and well completion techniques are particularly important.

A second concern relates to availability of submersible pumps for depths greater than those suitable for suction pumps. A variety of such devices are currently being evaluated, including pumps that are currently in use.

RCRA regulations require groundwater monitoring wells at permitted facilities and thus call for a large increase in the number of groundwater samples that will be taken at many sites. Given the 400 to 500 chemicals of potential interest, there is clearly a need to reduce the costs of analysis. Perhaps analytical screening techniques can be developed which will rapidly narrow the classes of chemicals that may be present in samples and thereby reduce the costs of detailed analyses.

STATISTICAL BASIS FOR SAMPLING PROGRAMS

During the past several years there has been growing interest in developing and improving the statistical bases for monitoring programs near hazardous waste sites. The application of statistics is becoming particularly important in legal proceedings which consider the validity of monitoring data. While statistical aspects of quality assurance programs are reasonably well developed, relatively little attention has been given to statistics in sampling design at hazardous waste sites.

Traditionally, "professional judgments" have been the primary determinant in the design of surface and subsurface sampling programs. Usually, these judgments are guided by expert knowledge of the geography and hydrogeology of the area, by some understanding of the mobility of the chemicals of concern, and by results of earlier sampling programs at the site. Also, the location of nearby residences and the likely environmental pathways to people in the neighborhood are usually considered. On occasion, statistical considerations have played a role in determining the locations and frequency of the sampling activities.

Similarly, professional judgments are a central element in interpreting monitoring results. For example, the contouring of concentration isopleths and the determination of the presence or absence of contamination patterns are generally based on subjective interpolations and extrapolations from values at individual sampling points displayed on maps or photographs. Pattern recognition and other types of statistical inferences have not become common components of data interpretation.

EPA's surface and subsurface monitoring activities at Love Canal in 1980 assessed the likely influence of sewers, roads, and other man-made structures on pollutant migration and earlier monitoring data collected by several organizations. Special attention was given to sampling along swales and other potential pollution pathways from the canal to residential areas and to sampling at residences which were alleged to have high levels of contamination or to be plagued with unexplained illnesses. The number of samples was limited principally by budgetary considerations, and the selection of specific sites was made by specialists on the scene who were limited to some extent by the lack of accessibility of a few preferred sites. The principal statistical concern was to take a sufficient number of samples in various geographical areas, including control areas, to permit comparisons among areas. As is the case in almost all sampling programs, budget constraints did not permit the large sample sizes which would have been considered optimal by statisticians.

The primary interpretations performed on the data that were collected were graphical and descriptive in nature. They consisted of attempts to identify patterns of contamination in the neighborhood near the canal and to determine contaminant gradients indicating pollutant migration away from the canal. Supplementary statistical approaches consisted of aggregating the data according to geographical areas and computing a variety of descriptive characteristics, (e.g., means, medians, ranges, frequency distributions). Limited efforts were directed to correlation analyses, principal components factor analyses, and cluster analyses. This combination of extensive subjective data analysis and limited statistical analysis provided the evidence as to the habitability of the Love Canal area.

Following the Love Canal experience, a technique relatively new to the environmental field—geostatistics—was introduced in soil monitoring efforts near lead smelters in Dallas and is now being considered for other applications. Since pollution plumes generally spread in a contiguous manner, samples taken close to one another should be more alike than samples taken farther apart. The contaminant levels in such closely associated samples are considered to be correlated variables which obey different statistical laws than random variables. For random variables, the best sampling schemes are random sampling designs. For correlated variables, the best approaches are systematic sampling schemes, such as sampling on a grid. Geostatistics is based on the concept of correlated variables.

For random variables, the variability of the pollution concentrations at the site and the acceptable uncertainty determine the sample size. The larger the variation at the site or the smaller the acceptable uncertainty, the more samples must be taken. For correlated variables, the relationships between samples, as well as the variability at the site and the acceptable uncertainty, determine the grid spacing. The closer the relationships, the fewer the samples that are required. Also, the tighter the grid, or the greater the number of samples, the smaller the uncertainty.

Computer programs are available for using preliminary monitoring data to estimate correlation factors for different spacings of sample points. These estimates can then be used as the basis for designing a more extensive sampling scheme. Sampling sites should be close enough together to assure that there is a correlation between adjacent sites attributable to the pollution plume but not too close to increase the cost of monitoring unnecessarily. The data that are acquired can then be manipulated to develop concentration isopleths based on objective statistical interpolations between sampling points. Also, isopleths of standard deviations can be generated to indicate those areas where the confidence in the data should be high or low. Subsequent sampling efforts presumably would be focused on those areas where the concentrations are high and/or the confidence is low.

In the years ahead, expanded uses of classical statistics and greater reliance on geostatistics and other innovative approaches for adding objectivity to sampling programs will be welcomed. However, professional judgments will continue to be critical to the design of every sampling program and to interpretation of monitoring data. In short, statistical approaches will guide and reinforce, but not supplant, professional judgments.

Passive Exposure Monitors

For many years the passive, personal dosimeter has provided much of the information used in estimating radiation exposures to workers at nuclear facilities and to other populations which might be exposed to radiation. Personal dosimeters are used for estimating both short-term exposures in areas suspected of being contaminated and long-term exposures in these and other areas to ensure that allowable levels for external radiation are not exceeded.

In recent years, the use of personal monitors for estimating chemical exposures has also received considerable attention. Both pumping devices and passive monitors that can be easily carried by individuals have been investigated by industrial hygienists, and such devices are now being used in some industrial settings.

Most recently, two types of more sensitive, pocket-size passive monitors for measuring the ambient levels of organic chemicals have been under development. The first type, patterned after an industrial hygiene model but with greater sensitivity and specificity, relies on charcoal trapping of the chemicals. Measurements have been made for a dozen or so chemicals in the 10-ppb range. Flow rates through the devices can be estimated, and the level of detection is lowered as exposure times – and hence total flow – increase. The second device relies on solid sorbents (e.g., Tenax®) for collection of the chemicals. Additional chemicals, including a few semivolatile organics, have been measured under a greater range of humidity conditions using this device.

If relatively inexpensive procedures for analyzing the cartridges of the passive monitors daily or weekly can be developed, these types of devices should have wide applicability. They should be useful both as personal dosimeters for workers at waste sites and as monitors affixed to telephone poles and other structures near waste sites.

Passive personal monitors may also help in estimating levels of indoor contamination where people spend most of their time. In particular, devices worn by small sample populations may provide a means for extrapolating from outdoor ambient air measurements of selected chemicals which are available to the actual exposures in a variety of environments over a 24-hour period.

USE OF AERIAL PHOTOGRAPHY TO ESTIMATE POPULATION PATTERNS

Another approach to exposure assessment involves relating pollutant levels that are measured at fixed locations (indoor and outdoor) in specific geographical areas to the population patterns in those areas. Frequently, waste sites are intermingled with industrial complexes, and exposure levels are aggregations of pollution discharges from a variety of sources. Determining ambient pollution levels in several media in a geographical area is often easier than determining the population that is in the area at any given time and thereby exposed to the chemicals. Tracking the movements of small population samples through the use of questionnaires or log books is one approach that has been traditionally used to correlate people and locations by the epidemiologists. This approach is costly, time consuming and difficult. A second technique relies on the use of aerial photography together with census data for determining population concentrations in specific areas at specific times and also population movements between the specific areas.

Using aerial photography and available monitoring data, a grid can be placed over the area of interest and pollution levels estimated for each grid cell. The next step is to estimate the population in each grid cell throughout the day. All houses, buildings, churches, schools, hospitals, shopping malls, and other features of interest are tallied for each grid cell. Census data provide information concerning the average number of people per housing unit as the basis for estimating the residential population within each grid cell. The working population is based on estimates of square units of work space in the cell. In one case, a reasonable estimate was 64 square meters per worker in light and heavy industrial buildings and 7 square meters in office buildings. The buildings can be identified in aerial photographs supplemented by data from other sources such as industrial directories. Facilities which attract shoppers and recreation seekers can be used to help estimate movements of local populations. Also, locations of particularly sensitive populations (e.g., school children, elderly, sick) are often tied to special types of facilities, such as schools, hospitals, and nursing homes.

Given the transient nature of pollution levels, particularly air pollution levels, and the changes in population patterns throughout the day, both pollution and population estimates can be presented in 8-hour increments:

midnight to 8 a.m., 8 a.m. to 4 p.m., 4 p.m. to midnight. In this manner, closer correlations can be made to actual exposures.

Historical photographs taken ten, twenty, or thirty years ago are frequently available in a variety of photographic archives and can be particularly useful in making some assumptions about retrospective exposures when the locations of the source of pollution and the exposed populations are uncertain. In addition to providing direct evidence concerning various activities, archival photography may be helpful in stimulating the memories of individuals who can provide other types of insights as to possible exposure problems.

CURRENT PRACTICE IN RELATING ENVIRONMENTAL MEASUREMENTS TO HEALTH IMPACTS

Two types of approaches for relating environmental measurements to health impacts are currently being pursued in assessments at hazardous waste sites. The most widely used approach defines a priori certain levels of contamination as unacceptable and requires corrective action if the monitoring program determines that these levels are exceeded. This approach is analogous to compliance monitoring for air and water pollution problems. The other approach is to obtain monitoring information and then determine whether the contamination levels, individually or in the aggregate, pose a health risk. Of course, in practice these two approaches are often considered together in the overall assessment.

Standards, Criteria, and Action Levels

Drinking water standards and advisory levels (for groundwater), water quality criteria (for surface water), and ambient air standards frequently provide the starting point in determining whether pollutants leaving a waste site pose a near-term exposure problem. Drinking water standards are also used as one of the factors in determining whether a waste should be classified as hazardous. Specifically, if a waste extract contains a contaminant for which a standard exists at a concentration exceeding the standard by a factor of 100 or more, the waste is classified as hazardous.

RCRA regulations call for corrective action if groundwater contamination "above background" is detected for selected contaminants, including some which are specified in permits on a site-by-site basis. The presumption is that in the long term such contamination could cause health or ecological problems unless the permittee can demonstrate otherwise. Consent decrees and other types of settlements also frequently specify contaminant levels of concern and call for monitoring programs to detect such levels.

The problems of widespread dioxin contamination in Missouri focused attention on the need for an action level to guide both the selection of monitoring methods adequate to measure the level and the implementation of the cleanup program. The action level of 1 ppb in soil established by the Centers for Disease Control presumably reflects an assessment of potential health risks to humans.

Classical Approach to Risk Assessment

We have had little success to date in moving from monitoring data to exposure assessments to risk assessments at hazardous waste sites in a formalized sense. The multiplicity of chemical mixtures involved, the uncertainties associated with current and future multiple environmental pathways to human populations, and the time lags from exposure to manifestation of chronic health or ecological effects have greatly complicated the use of models for risk assessment. Still, these models are currently available tools in helping to use environmental monitoring data and helping to guide collection of future data.

MONITORING DATA TO SUPPORT HEALTH STUDIES

Anecdotal reports of symptoms of adverse health impacts experienced by segments of the population near hazardous waste sites suggest that a wide range of both acute and chronic effects should be of interest. Authoritative monitoring data is a critical aspect of health studies that attempt to relate such health effects to exposures resulting from chemical discharges from waste sites.

Epidemiological studies are usually initiated in those cases when the impacted population is sufficiently large and/or the effect is sufficiently unique to permit meaningful statistical analyses. Case-control and cohort studies are probably the most appropriate in addressing hazardous waste problems. In these, estimates of past exposure levels are particularly important.

When the population is not large and the alleged effects not unique, the sample size may preclude a meaningful interpretation. Nevertheless, qualitative assessments as to possible cause and effect relationships are still very important. Such judgments presumably can be greatly enhanced by sound data on likely exposure levels throughout the impacted population.

The types of monitoring data that are currently being collected and the scope of collection activities vary considerably from site to site. As the newer monitoring techniques described earlier in this presentation find greater use, the types of available monitoring data will expand, with the health researcher a primary beneficiary of this enlarged database.

CHAPTER 4

Assessing Pathways to Human Populations

Julian B. Andelman

INTRODUCTION

To determine whether a hazardous waste site is likely to have caused or will cause adverse human health effects, it is necessary to assess the pathways along which the hazardous constituents move and ultimately cause human exposures. This is one of the principal components of "exposure assessment" which in turn is an essential part of an environmental epidemiological study or indeed any hazard assessment of a waste site. Without such a clear understanding of these pathways and their significance, the hazards cannot be fully understood, nor can the wisest and most cost-effective remediation and other mitigation actions be undertaken.

The purpose of this chapter is to consider some of the important processes that affect these pathways, not so much to instruct the environmental scientist as to provide an orientation to the epidemiologists and other health scientists undertaking studies or evaluating the hazards of potentially exposed populations.

J. B. Andelman and D. W. Underhill (Editors), *Health Effects from Hazardous Waste Sites*
© 1987 Lewis Publishers, Inc., Chelsea, Michigan – Printed in U.S.A.

Table 1. Chemical Species of Arsenic

Oxidation State	Formula	Name
+5	As_2O_5	arsenic pentoxide
+5	H_3AsO_4	arsenic acid
+5	AsO_4^{3-}	arsenate
+3	As_2O_3	arsenic trioxide
+3	H_3AsO_3	arsonic acid
+3	AsO_3^{3-}	arsenite
+3	$(CH_3)AsO(OH)_2$	methyl arsonic acid
+1	$(CH_3)_2AsO(OH)$	dimethyl arsinic acid, cacodylic acid
−3	AsH_3	arsine
−3	$(CH_3)_3As$	trimethyl arsine

PHYSICOCHEMICAL ASPECTS

Physicochemical properties of chemicals affect their movements and fates in the environment, the resulting human exposures, and ultimately the potential toxicological impacts. One such important characteristic is the oxidation state. Arsenic is an interesting example of a toxic chemical with several oxidation states with likely different toxicities, important examples of which are shown in Table 1. The +5 state normally dominates in well-aerated waters in the form of arsenate (AsO_4^{3-}), while the +3 form, arsenite (AsO_3^{3-}), is the predominant species in anoxic waters.[1] There is a longstanding U.S. drinking water standard for arsenic of 0.05 mg/L, but rarely, at least for regulatory purposes, is any attempt made at "speciating" the chemical forms in public water supplies. As_2O_3 is generally considered to be the form associated with lung cancer in inhalation industrial exposures, such as at copper smelters.[2] Whether the +5 or +3 form is the more likely one as an agent of skin cancer is not certain; there is also some evidence of inter-conversion among these two oxidation states in some animals and humans.[3] The −3 state is manifested in arsine (AsH_3) which is highly toxic and volatile and, therefore, behaves in a fashion clearly different from these other two forms. It is stable only in a highly reduced environment. Finally, as shown in Table 1, there are "organic" forms of arsenic, i.e., with organic substituents covalently bonded to the arsenic, such as dimethyl arsinic acid or cacodylic acid. This has been used as a medicant and is applied in agriculture as an herbicide, desiccant, and defoliant.[4] There is also evidence that it is biologically synthesized.[5] Thus, the complexity of the arsenic cycle in nature should be considered when there is a potential arsenic source in the environment. Other chemicals as well have highly different behavior and effects associated with such speciation, notably mercury, where the inorganic form is quite different in toxicity and behavior from the organic ones, such as dimethyl mercury.[6] The transformations between these forms in the environment have been studied extensively. Initially, in the well-known

Table 2.   Aqueous Solubility, Vapor Pressure, and Henry's Law Constants For Several Organic Compounds at 25°C[8]

Compound	Aqueous solubility mg/L	Vapor Pressure mm Hg	H atm m^3/mol
benzene	1780	95.2	5.5×10^{-3}
toluene	515	28.4	6.7×10^{-3}
naphthalene	33	0.23	1.2×10^{-3}
biphenyl	7.5	0.057	1.6×10^{-3}

Minamata Bay "outbreak" of mercury poisoning, there was considerable confusion because the contaminating source was inorganic mercury while the symptomatology was typical of organic mercury poisoning. Until the essential link of environmental transformation between the two forms was established, the pathway remained problematic.[7]

Solubility of chemicals in water and their vapor pressures are two important interactive characteristics that affect their movement through water and their transfer into air. As shown in Table 2, vapor pressures of anthropogenic chemicals can vary widely, as can their solubilities in water. If there is a pure source of that chemical at a waste site, for example, a "pool" of the liquid at or near the soil-air interface, clearly the chemical with the higher vapor pressure will tend to be transported in greater quantities by volatilization into the air and subsequent air movement. Similarly, to the extent that it is more soluble in water, the chemical will tend to move into it more readily, such as by leaching and subsequent transport through an aquifer. However, what is appreciated, perhaps to a lesser extent, is that the interaction between aqueous solubility and vapor pressure is a more important predictor of volatilization from water. This is expressed by the Henry's law constant, H, which quantitates the equilibrium between the partial pressure of the chemical in air, P_{atm}, and its aqueous concentration, C_{aq}, namely:

$$H = P_{atm}/C_{aq} \tag{1}$$

As shown in Table 2, chemicals with very different aqueous solubilities and vapor pressures can have similar Henry's law constants and thus be expected to volatilize from water to the same extent if they are present at similar aqueous concentrations. Also, the volatilization of such constituents from indoor uses of contaminated water may constitute an important inhalation exposure, as discussed by Harris et al.[9] in Chapter 12.

As a further illustration of this effect, we studied a community near Pittsburgh, namely Valencia, Pennsylvania, a residential area of perhaps one hundred homes using water supplied by individual wells. There were a variety of complaints of illness thought to be associated with water use. In our survey of these homes in which the water was sampled, about 10% had measurable benzene concentrations (above 0.1 μg/L), with several in the

Table 3.   Benzene in Selected Well Water Samples in Valencia, PA, 1983[10]

Well	Benzene concentrations μg/L (ppb)	
	March 5	March 26
3	87	14
4	219	77
7	265	120
15	24	8
37	1	ND[a]
66	ND	ND
68	213	74

[a]Below detection limit of 0.1 ppb.

range of 200 to 300 μg/L.[10] A few results from this study are shown in Table 3. A few samples of air from the homes were analyzed and found to contain benzene, indicating exposure through this volatilization route. For one home with a water concentration of 214 μg/L, the air concentration was 122 μg/m^3. For reference, a seven-day SNARL (suggested no adverse response level) for benzene of 12.6 mg/L in water has been recommended by the National Academy of Sciences (NAS).[11] At the same time, the American Conference of Governmental Industrial Hygienists (ACGIH) time-weighted threshold limit value for an eight-hour, five-day occupational exposure is 30 mg/m^3 in air.[12] Considering these domestic exposures could have been occurring for a long time, they should be of concern. Also the measured air concentrations are an indication of the potential importance of this inhalation-volatilization route of exposure. Subsequent studies by this author have confirmed its importance in domestic shower and bath systems.[13] Such volatilization is also an important mechanism for potential air exposures from contaminated surface waters, as well as in the vicinity of a hazardous waste site.

Another aspect of chemical behavior that should be considered in assessing pathways is chemical reactivity. Thus, although a chemical may be leached into an aquifer, whether it ultimately results in an exposure will depend on whether it maintains its chemical integrity. For example, halogenated methanes can hydrolyze in aqueous systems to form methyl alcohol and halogen acids. However, their rate of hydrolysis is highly variable, depending on such factors as pH and chemical structure, as shown in Table 4. Not unexpectedly, the greater the number of halogen atoms, the greater is the stability (e.g., carbon tetrachloride > chloroform > methylene chloride > methyl chloride). Also, stability decreases in the order of chloride, bromide, iodide. At the extremes, the eight-hour half-life of methyl iodide in this hydrolysis reaction is in sharp contrast to the 70,000-year half-life of carbon tetrachloride. This is only one example of possible chemical reactivity that should be considered, particularly when the chemical exposures from a waste site can occur at a distance and are associated with a time

Table 4. Estimated Half-Lives of Halogenated Methanes in Hydrolysis Reactions in Aqueous Solution[14]

Compound	$t_{1/2\ hours}$	$t_{1/2\ years}$	Temp., °C	pH
methyl chloride	10^4	1	25	4–10
methyl bromide	480	5×10^{-2}	25	4–10
methyl iodide	8	10^{-3}	69	7
methylene chloride	6×10^6	7×10^2	–	–
	1.5×10^4	1.5	25	–
chloroform	3×10^8	3×10^4	25	4
	3×10^7	4×10^3	25	7
	3×10^4	4	25	10
	1.3×10^4	1.3	25	–
carbon tetrachloride	6×10^8	7×10^4	–	1–7

delay. By not considering these degradative processes, one could either overestimate the possible exposures of the parent compound or, alternatively, not be aware of those from the products of the reactions.

Another group of compounds which has been the subject of considerable attention relative to exposure by both the water and air routes is the polycyclic aromatic hydrocarbons (PAH). Although previously studied widely as air contaminants, starting in the late 1960s attention became focused on their presence and behavior in natural and treated water systems.[15] Potential adverse health effects center on their varying range of carcinogenicities, with 3,4-benzopyrene (BP) exemplifying one of the more prevalent and carcinogenic forms. Although as a group of compounds they are relatively insoluble, they nevertheless are ubiquitous in polluted waters, often at concentrations substantially higher than expected based on water solubility.[15] This is primarily because they adsorb very effectively onto a variety of surfaces, including the suspended particulate and colloidal matter found in most natural, fresh surface waters.[15] Thus, their presence is explicable, as is their ready removal in conventional water treatment designed to remove particulate matter. An assessment of their behavior in water systems that does not consider these phenomena could be highly inaccurate in its estimation of human exposures.

Often the ability of relatively hydrophobic organic compounds to sorb (a term to encompass either adsorption or absorption) depends on their relative solubilities in organic solvents, i.e., in octanol vs that in water.[16] This is usually expressed as the octanol-water partition coefficient, P, which can vary over several orders of magnitude, depending on the compound, as shown in Table 5. The extent of sorption onto soils is often related to this partition coefficient and the fraction of organic matter on the soil. Another manifestation of the partitioning process and, hence, the partition coefficient is its correlation with bioconcentration, i.e., the process involving extraction from water into fatty and similar matter of aquatic organisms and other aquatic biota. The bioconcentration factor, BF, has been shown

Table 5. Partition Coefficient and Bioconcentration Factor For Several Chemicals[16]

Chemical	log P	log BF
1,1,2,2-tetrachloroethylene	2.88	1.59
carbon tetrachloride	2.64	1.24
p-dichlorobenzene	3.38	2.38
diphenyl oxide	4.20	2.29
diphenyl	4.09	2.64
2-biphenyl phenyl ether	5.55	2.74
hexachlorobenzene	6.18	3.89
2,2',4,4'-tetrachlorodiphenyl	7.62	4.09

to correlate very well with the partition coefficient, as shown for typical values in fish in Table 5.[16] Thus one can expect that an organic chemical with a high P value, leaching from a hazardous waste site into an aquifer and ultimately reaching a surface water, has the potential for bioconcentrating in fish. This is an additional pathway that can result in human exposure, possibly even at higher levels than that involving the direct ingestion of the contaminated water.

MODELING PATHWAYS

A number of attempts have been made to model the movement of chemicals in the environment so as to ultimately assess human exposure. Some of these can be quite complex, but nevertheless can be of assistance in determining the possible impacts of the chemicals emanating from a point source. For example, Neely et al.[17] reported the results of their modeling of a chloroform spill in the Mississippi River, following which the measured variation in concentration downstream for considerable distances was interpreted in terms of movement of chloroform among "compartments" of the river and into the atmosphere. Their model fitted to the measured concentrations is shown in Figure 1. The authors concluded that there was significant transport into the bulk aqueous compartment of the river from the chloroform pool at the bottom, as well as volatilization into the air. At about ten hours, the chloroform reached a very high concentration (about 300 $\mu g/L$) at a point downstream from the spill. In comparison, the typical concentration of chloroform in U.S. public water supplies is in the range of 15 to 30 $\mu g/L$. This study indicates that the behavior of the chemical following the spill and its movement into the aqueous and atmospheric phases can be understood in terms of known physicochemical phenomena. This behavior of chemicals should be considered in estimating human exposures that can ensue from point source emissions into flowing surface waters.

As complex as surface water modeling can be, groundwater modeling of the movement of anthropogenic chemicals can be even more so and is

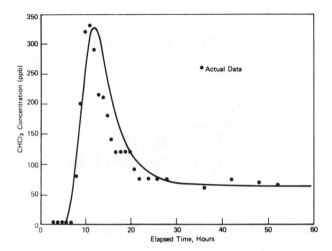

Figure 1. Concentration time profile of chloroform (16.3 mi from point of addition). (Reprinted with permission from *Environmental Science and Technology* 10(1), January 1976. Copyright 1976 American Chemical Society.)

currently at a more primitive state of understanding. A variety of questions relative to assessing pathways and ultimately human exposures are appropriately addressed by such models. If the chemical of interest were "conservative," that is did not react in the groundwater system nor interact with sediments or the minerals of an aquifer, then one would expect to find, after a certain period of time at a monitoring site remote from the point source leaching into the aquifer from a hazardous waste site, that the chemical "appeared" and might even start to increase in concentration. If the chemical front moves along at the same rate as the water in the aquifer, there is "plug" flow. If the point source constituted a sudden release that also terminated suddenly, then with such plug flow at a distant point, the concentration would also decrease suddenly with time after its buildup. With a continuous source, however, the conservative constituent would tend to maintain a constant concentration at the monitoring point in a plug flow situation following the arrival of the front.

The behavior of the chemical in the aquifer can, however, become much more complex and include such phenomena as longitudinal and transverse dispersion or spreading, sorption and other interactions with the lithographic structure of the aquifer, and chemical and microbiologically mediated degradation. Some of these phenomena have been reviewed and studied by McCarty et al.[18] They noted that, in an anoxic aquifer, there can be sufficient food sources for microorganisms so that, in addition to chemical transformations, biologically mediated transformations can occur as well. However, if the waste site leachate or other possible chemical inputs into the

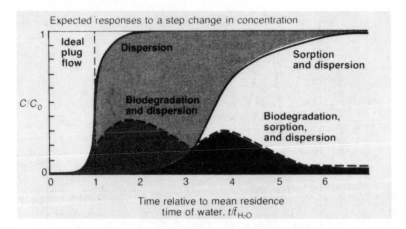

Figure 2.   Effects of dispersion, sorption, and biodecomposition on the time change in concentration of an organic compound at the aquifer observation well. (Reprinted with permission from *Environmental Science and Technology* 15(1), January 1981. Copyright (1981) American Chemical Society.

aquifer are below a critical level, then such biodegradation might very well not be important.

McCarty et al. considered the likely impact of adsorption on the concentrations of several organic contaminants "downstream" in an aquifer,[18] noting that their arrival at a distant point can be delayed by adsorption processes, analogous to such delays that occur in a laboratory chromatographic column. The effects of these adsorption processes, dispersion, and biodegradation acting either separately or in concert are shown in Figure 2. This describes how the concentration of an organic chemical contaminant from a continuous source can vary with time at an observation well as water flows past it. It shows that with ideal plug flow the concentration rises suddenly as the front reaches the well. In contrast, with dispersion, there is a more gradual rise. With the incorporation of the sorption process, there is a considerable delay and a more gradual rise. Nevertheless, after the adsorbing surfaces of the upstream structures become saturated, the concentration at the monitoring well reaches the level that is attained in plug flow. However, with sufficient biodegradation a peak concentration can be observed, with a final leveling off to a lower steady-state concentration at some later time.

McCarty et al. compared the actual delays in movement, or retention times, of chloroform, chlorobenzene, and 1,4-dichlorobenzene injected into a reclamation system with the expected values based on octanol-water sorption coefficients.[18] The comparisons are shown in Table 6 and indicate reasonably good agreement. As discussed earlier, the degree of coating of the sorbing minerals with organic matter plays an important role in estab-

Table 6. Comparison Between Field-Measured and Predicted Relative Retention Times For Various Organic Compounds[18]

| Compound | P | Relative retention time | |
		Measured	Predicted
chloroform	93	5	6
chlorobenzene	692	36	41
1,4-dichlorobenzene	2400	>200	140

lishing the extent of sorption, as does the octanol-water coefficient. McCarty et al. noted that in their studies, the retention is quantitatively related to these factors as long as the organic fraction of the adsorbing minerals is at least 0.1%. Such studies and predictions provide a useful framework for understanding the movement of chemicals leaching from a waste site into and transport through an aquifer. They demonstrate the likely complexities of these processes and the need for both caution and considerable expertise in assessing this groundwater pathway and the eventual possible human exposures that might ensue.

Finally, it should be noted that a variety of computerized models are being and have been developed to take into account these various physico-chemical and biological phenomena to assess the stability, movement, and fate of chemicals in aqueous, as well as air, media. Although such models should always be used with caution and verified for the particular area of study, they can be of assistance in at least providing a range of estimates of exposures that could result from a given pathway. An example of such a model for synthetic organic chemicals in aquatic ecosystems is the Exposure Analysis Modeling System (EXAMS) developed by the U.S. Environmental Protection Agency.[19] This model determines the persistence, fate, and exposures that can occur for various chemicals using information about their chemistry and physicochemical characteristics. It is a multicompartment model designed to assist in hazard evaluation. Although it cannot be considered as a substitute for an analysis using site-specific hydrogeological and related information, it can aid the scientists assessing pathways, exposures, and ultimately potential effects from hazardous waste sites by providing some basis for the range of expected exposures. Whether the assessment of pathways uses such computer models or other methodologies with appropriate consideration of physicochemical behavior, the epidemiological studies and/or health risk analyses need such inputs to be effective. The alternative is to achieve at best only a qualitative understanding of the likely impacts with a possibly significant deficiency in the ultimate assessment and basis for mitigating the hazards associated with a waste site.

REFERENCES

1. Irgolic, K. J., Stockton, R. A. and Chakraborti, D. Determination of Arsenic and Arsenic Compounds in Water Supplies. In Lederer, W. H. and Fensterheim, R. J., Eds. *Arsenic: Industrial, Biomedical, Environmental Perspectives.* Van Nostrand Reinhold Company, New York, ch. 22, 1983.
2. Enterline, P. E. and Marsh, G. M. Mortality Among Workers Exposed to Arsenic and Other Substances in a Copper Smelter. In Lederer, W. H. and Fensterheim, R. J., Eds. *Arsenic: Industrial, Biomedical, Environmental Perspectives.* Van Nostrand Reinhold Company, New York, ch. 19, 1983.
3. Peoples, S. A. The Metabolism of Arsenic in Man and Animals. In Lederer, W. H. and Fensterheim, R. J., Eds. *Arsenic: Industrial, Biomedical, Environmental Perspectives.* Van Nostrand Reinhold Company, New York, ch. 11, 1983.
4. Abernathy, J. R. Role of Arsenical Chemicals in Agriculture. In Lederer, W. H. and Fensterheim, R. J., Eds. *Arsenic: Industrial, Biomedical, Environmental Perspectives.* Van Nostrand Reinhold Company, New York, ch. 5, 1983.
5. Ferguson, J. F. and Gavis, J. A Review of the Arsenic Cycle in Natural Waters. *Water Res.* 6:1259–1274, 1972.
6. National Academy of Sciences (NAS). *An Assessment of Mercury in the Environment.* National Academy Press, Washington, DC, p. 4, 1978.
7. National Academy of Sciences. *An Assessment of Mercury in the Environment.* National Academy Press, Washington, DC, ch. 3, 1978.
8. Mackay, D. and Leinonen, P. Rate of Evaporation of Low-Solubility Contaminants from Water Bodies to the Atmosphere. *Environ. Sci. Technol. 9:1178–1180, 1975.*
9. Harris, R. H., Rodricks, J. V., Clark, C. S. and Papadopulos, S. S. Adverse Health Effects at a Tennessee Hazardous Waste Disposal Site. This volume.
10. Fadzen, S. Petrochemical Contamination of the Valencia Groundwater Supply. M.Sc. (Hygiene) Thesis, Graduate School of Public Health, University of Pittsburgh, 1983.
11. National Academy of Sciences (NAS). *Drinking Water and Health. Volume 3,* National Academy Press, Washington, DC, 1980.
12. American Conference of Governmental Industrial Hygienists (ACGIH). *Threshold Limit Values for Chemical Substances in the Work Environment Adopted bv ACGIH for 1983–84.* ACGIH, Cincinnati, OH, 1983.
13. Andelman, J. B. Inhalation Exposure in the Home to Volatile Organic Contaminants of Drinking Water. *Sci. Total Environ.* 47:443–460, 1985.
14. National Academy of Sciences (NAS). *Chloroform, Carbon Tetrachloride and Other Halomethanes: An Environmental Assessment.* National Academy Press, Washington, DC, 1978.
15. Andelman, J. B. and Suess, M. J. Polynuclear Aromatic Hydrocar-

bons in the Water Environment. *Bull. World Health Org.* 43:479–508, 1970.

16. Neely, W. B. *Chemicals in the Environment.* Marcel Dekker, Inc., New York, 1980.

17. Neely, W. B., Blau, G. E., and Alfrey, Jr., T. Mathematical Models Predict Concentration-Time Profiles Resulting from Chemical Spill in a River. *Environ. Sci. Technol.* 10:72–76, 1976.

18. McCarty, P. L., Reinhard, M., and Rittmann, B. E. Trace Organics in Groundwater. *Environ. Sci. Technol.* 15:40–49, 1981.

19. Burns, L. A., Cline, D. M., and Lassiter, R. R. *Exposure Analysis Modeling System (EXAMS): User Manual and System Documentation.* U. S. Environmental Protection Agency, EPA-600/3-82-023, Athens, GA, 1982.

CHAPTER 5

Biological Methods of Defining Human Exposures

Laszlo Magos

INTRODUCTION

The evaluation of human exposure to toxic substances is a complex process which includes both environmental and biological methods. Biological monitoring, according to the definition of Zielhuis[1] is "the measurement of internal exposure through the analysis of a biological specimen" and "must clearly be distinguished from the evaluation of health effects" like screening for bladder cancer, enzyme changes, or chromosomal abnormalities. Though monitoring means continuous measurement and the biological assessment of exposure is carried out at only one or more selected points in time, biological monitoring is not a misnomer. Biological monitoring is expected to give an index of exposure, which would otherwise only be given by prolonged and complex environmental assessment. The longer is the clearance half-time of the toxic substance, the longer is the exposure period which influences the biological indicators of exposure, and the longer is the postexposure period which permits the use of biological indicators. When exposure is high enough to produce metabolic or other effects, which persist

J. B. Andelman and D. W. Underhill (Editors), *Health Effects from Hazardous Waste Sites*
© 1987 Lewis Publishers, Inc., Chelsea, Michigan – Printed in U.S.A.

Table 1. Biological Methods Used For the Assessment of Exposure to Toxic Substances

Exposure test		Examples for lead
direct	toxic substances and/or their metabolites in biological specimens	lead in blood
indirect	indicators of metabolic interference in biological specimens	ALAD activity in blood, aminolevulinic acid in urine
clinical	indicators of functional or morphological injury	nerve conduction velocity, aminoaciduria
retrospective	morbidity and mortality	chronic nephropathy

longer than the time limit for biological monitoring, these effects may be used as indicators of unacceptably high exposure.

The concept and practice of biological monitoring was developed in the field of industrial toxicology, because, as Elkins,[2] an early advocate of biochemical tests in the evaluation of toxic hazards, pointed out: "the concentration of a toxic agent in body tissues or excreta often gives a more accurate measure of the worker's exposure than can be obtained by any tests on his environment." The origin of biochemical exposure tests explains why most of the examples listed later are from industrial toxicology. However, irrespective of their origin, all can judiciously be applied to the evaluation of hazards presented from any source, including the dumping of toxic waste. This judicious approach is imperative because nonoccupational exposure does not follow the regular exposure pattern of workers and the sensitivity of a population which includes individuals above and below the working age and pregnant women, may be different from the sensitivity of workers.

THE CATEGORIES OF BIOLOGICAL METHODS USED FOR EXPOSURE EVALUATION

Table 1 lists the different biological methods used for exposure evaluation. Of the four categories, only direct tests satisfy the definition of Zielhuis.[1] However, when a biochemical abnormality is present below the accepted tolerable limit of exposure and increases with exposure, its measurement can be accepted as a biological monitoring method. Zielhuis[3] gives as examples the changes in the activity of ALAD or in the concentration of zinc-protoporphyrin.[1] Another example is the estimation of methemoglobin in workers exposed to aromatic amino and nitro compounds.[4]

The demarcation between the different test categories is not sharp. For example the estimation of carbon monoxide (CO) in blood is clearly a direct exposure test. However, blood CO is associated with hemoglobin and the

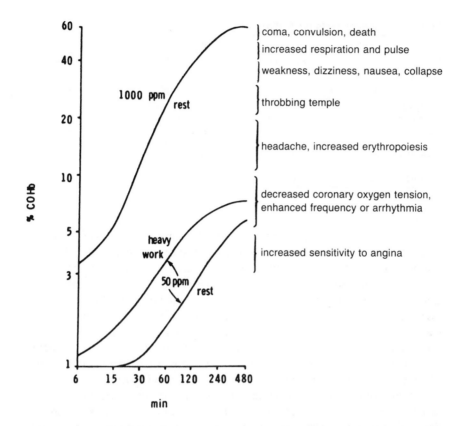

Figure 1. Expected COHb values under three conditions according to Peterson and Stewart[44] and the correlation between % COHb and the toxic effects of CO.[45]

estimation of this abnormal blood pigment, carboxyhemoglobin (COHb), fits into the category of indirect tests. In addition to being both a direct and an indirect test, the estimation of COHb is also a diagnostic test because the impairment of oxygen transport through the formation of COHb is the essential step responsible for the toxic effects of CO. Figure 1 shows that these toxic effects range from the enhancement of irregular heart beats between 5 to 10% COHb to death around 60%. The correlation between the saturation of hemoglobin with CO and toxic impairments is so good that the onset of toxic signs can be predicted from the concentration of COHb, and similarly the degree of Hb saturation can be predicted from the severity of intoxication. Figure 1 also demonstrates that the saturation of Hb does not depend only on the atmospheric concentration of CO, but also on exposure time and on the work load.

Accurately defining exposure may require a battery of tests. Table 2 lists

Table 2. Exposure Tests for Aniline[4]

Test	Exposure		
	Acceptable	Moderate	Severe
In Urine:			
p-aminophenol	<7.0 mg/L	>7.0 mg/L	>7.0 mg/L
In Blood:			
Hb concentration	no effect	no effect	decreasing
methemoglobin	normal	elevated but below 1.4 g%	>1.4 g% >0.35 g%
verdoglobin	normal	elevated but below 0.35 g%	
Heinz bodies	none	<1%	>1.0%

exposure tests for aniline. The estimation of p-aminophenol urine, which is a direct exposure test along with the indirect test of methemoglobin determination, can define current exposure. The presence of Heinz bodies in red blood cells and an irreversibly oxidized form of hemoglobin, called verdoglobin, and especially a decline in Hb concentration are the evidence of unacceptably high exposure for a prolonged period. When exposure declined a few days before sampling, Heinz bodies, verdoglobinemia, and anemia may be present without any increase in urinary p-aminophenol excretion or methemoglobin concentration. Conversely, increased p-aminophenol excretion and methemoglobin concentration, without any shift in the other values, indicate a sudden increase in current exposure.

These tests, with the exception of urinary p-aminophenol excretion, are applicable to many aromatic amino and nitro compounds, like trinitrotoluene.[5] Naturally the limits shown on Figure 2 do not apply to those compounds, which like benzidine or 2-naphthylamine, can produce bladder cancer. When the manufacture and use of such compounds are not banned, any test which indicates an effect, even a slight methemoglobinemia, is a warning sign of unacceptably high exposure. However, the final verdict on exposure is given by more clinically inclined tests and after a long exposure time, e.g., when the cytological analysis of urine may already reveal the presence of bladder papilloma or cancer. This is only one example when the analysis of mortality and morbidity data is used to define exposure retrospectively through its health effects.

Though according to Zielhuis[1] every biological monitoring program is an epidemiological study, i.e., a study of the relationship between internal exposure (dependent variable) and the conditions of exposure (independent variable), the term "epidemiological evaluation" in industrial and environmental toxicology usually refers to studies in which the dependent variable is either a signal of risk, e.g., alkylation of DNA or Hb, or the incidence of disease(s) which can be associated with a particular exposure. When environmental or internal exposure history is defined and quantified, the associ-

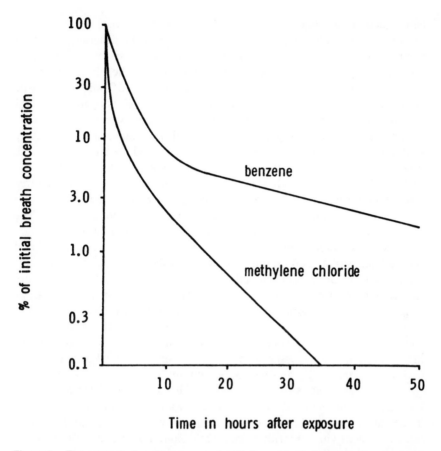

Figure 2. The concentration of benzene and methylene chloride in breath after exposure. Curve showing the decline of benzene concentration in breath was redrawn from Berlin et al.[19] and the methylene chloride exhalation curve from Stewart et al.[46]

ation between exposure and its effects can help to establish the upper limit of acceptable exposure.

THE IMPORTANCE OF MONITORING TOXIC EFFECTS FOR EXPOSURE ASSESSMENT

In industry, when the nature of exposure, the effects of exposure, and the biological and environmental threshold limits of toxic exposure are known, biological and/or environmental monitoring programs can be carried out routinely according to the most suitable design. However, even in industry, periodic medical examination, the analysis of personal health records, or new toxicological data occasionally reveal a need to screen workers for a certain type of health effect. While the exposure of the general population to toxic agents can be similarly monitored, in many cases biological monitoring, if implemented at all, is only secondary to the identification of exposure by its effects. The following examples taken from the fields of industrial and environmental toxicology demonstrate that environmental monitoring is often inadequate for the evaluation of exposure.

The first example is of gross environmental mismanagement.[6] A factory started to burn industrial waste in an open space, and the fire spread to the rubble-filled shaft of an abandoned coal mine. Carbon monoxide from the shaft reached the nest of adjacent bungalows and caused severe and, in three cases, fatal intoxication. The responsible agent and the degree of exposure were identified in the course of clinical and postmortem investigations. The source of CO was revealed only after the coroner ordered an inquiry.

The second example concerns investigations implemented after a national geographical survey had established that the village of Shipham in Somerset, United Kingdom had been built on cadmium-contaminated waste material derived from a zinc mine. A noninvasive neutron activation analysis found 11 ± 2 ppm cadmium in the liver of villagers, which was five times more than in a control group, and two orders less than concentrations found in industrial workers exposed to cadmium.[7] This biological monitoring program did not reassure people, and therefore clinical tests were carried out and morbidity and mortality data analyzed.[8-9]

The third example is the case of vinyl chloride. Maltoni[10] reported that vinyl chloride in rats produced angiosarcoma and other tumors. The report immediately prompted epidemiological surveys which confirmed the carcinogenic effect of vinyl chloride in man[11] and led to the revision of the threshold limit value (TLV).

The fourth example is the controversial subject of lead effects on the intellectual and behavioral development of children. The controversy is not about the quantification of exposure, but about the level of exposure which affects the development of children. Therefore, studies on environmental lead usually include tests beyond the scope of biological monitoring.

These examples demonstrate that the evaluation of exposure must occasionally include tests for both the measurement of exposure and the effects of exposure because only the absence or presence of effects can show whether exposure is or was acceptable. First, TLVs are based on the assumption that they assure safety, but this assumption always includes a nonstatic factor called "present knowledge." Second, when exposure is discontinued, past exposure may be judged only by the use of clinical methods or mortality and morbidity data. The fact that toxic effects are present only in the case of unacceptably high exposure is not an argument against the use of effects in the evaluation of exposure. Such an exposure is not the choice of the investigator, but it forces the investigator to use methods which not only estimate the degree of exposure, but also the frequency, nature, and degree of damage(s) inflicted on the exposed population. Luckily, in the majority of cases, direct and indirect tests, which measure exposure below the toxic range, satisfy the requirement of exposure assessment.

EXPOSURE TESTS

Direct Exposure Tests

The biological specimens for direct exposure tests can be blood, urine, breath, hair, teeth, and even a whole organ, as in the noninvasive neutron activation analysis of liver and kidneys for cadmium.[12]

Table 3 gives examples of direct exposure tests in blood with indicator values for excessive exposure. Sampling time, as with other specimens, depends on the clearance half-life of the toxic agent. The long half-life of methylmercury, lead, and cadmium assures that fluctuation of daily exposure periods—be it industrial, polluted urban air, or food—does not result in any appreciable fluctuation in the blood concentration of these metals. However, the blood concentration of acetone declines with a 6-hr half-life after the cessation of exposure,[13] and, therefore, the time of sampling is important. The half-life of methylene chloride is less than 30 minutes so it is easier to estimate its CO metabolite in the form of COHb.[14] Thus, COHb estimation, which is an exposure test for carbon monoxide, can also be used for the evaluation of methylene chloride exposure, and in both cases 5% COHb is accepted as a biological upper limit of acceptable exposure.[15]

Table 4 lists exposure tests in urine. The short clearance half-lives of solvents require an end-of-the-shift collection, while with metals, the exposure-dependent day-to-day fluctuation is not important. In the case of styrene, phenylglyoxylic acid gives better correlation with exposure than mandelic acid.[16] Some of the indicators are normal urinary constituents, and, therefore, their urinary excretion correlates with the exposure only above their background value. However, while the background level of urinary hippuric acid excretion can be as high as 1600 mg/L, phenol in the urine of nonexposed persons is not more than 30 mg/L.[16]

Table 5 lists direct exposure tests in breath. An important factor in gas exchange between alveolar air and alveolar capillaries is the partition coefficient between blood and air. Solvents with relatively low solubility in blood, e.g., methylchloroform, have a slower pulmonary uptake and achieve equilibrium between alveolar air and blood sooner than solvents, e.g., trichloroethylene, with high solubility. However, steady state also depends on their metabolism and on their extraction from blood by other tissues. When their initial blood concentrations are equal, the clearance of toluene is ten times faster than the clearance of styrene.[17] The high solubility of toluene in fat

Table 3.   Direct Exposure Tests in Blood

Substance	Indicator (if different from substance)	Indication of Excessive Exposure, >mg/L		References
		In Whole Blood	In Plasma	
acetone		200.0		Baselt, 1980[15]
acetonitrile	cyanide	0.1		Baselt, 1980[15]
acrylonitrile	thiocyanate		20.0	Baselt, 1980[15]
aldrin	dieldrin		0.005	Baselt, 1980[15]
cadmium		0.01		WHO Study Group, 1980[46]
hexachlorobenzene		0.05–0.1		Baselt, 1980[15]
lead		0.3–0.4		WHO Study Group, 1980[46]
methylmercury	Hg	0.2		WHO Task Group, 1976[47]

Table 4. Direct Exposure Tests in Urine

Substance	Indicator (if different from substance)	Indication of Excessive Exposure, > mg/L[a]	References
acetone		270.0[b]	Baselt, 1980[15]
arsenic		1.0	Baselt, 1980[15]
benzene	phenol	75.0[b]	Baselt, 1980[15]
cadmium		0.015	Baselt, 1980[15]
chromium		0.04	Baselt, 1980[15]
ethylbenzene	mandelic acid	2000.0[b]	Baselt, 1980[15]
mercury (Hg°)		0.08	WHO Study Group, 1980[46]
selenium		0.1	Baselt, 1980[15]
styrene	{ mandelic acid	3000.0[b]	Baselt, 1980[15]
	{ phenylglyoxylic acid	560.0	Baselt, 1980[15]
tetraethyllead	lead	0.11	Baselt, 1980[15]
thallium		0.005	Baselt, 1980[15]
toluene	hippuric acid	2800.0[b]	Baselt, 1980[15]
trichloroethylene	trichloro compounds	400.0	Lauwerys, 1975[16]
xylene	m-hippuric acid	2630.0[b]	Baselt, 1980[15]

[a]Values corrected to 1.024 specific gravity.
[b]End of shift collection.

Table 5. Direct Exposure Tests in Breath

Substance (and exposure, ppm)	End Shift Concentration in Breath, ppm	References
acetone (1000)	250	Baselt, 1980[15]
benzene (10)	0.2	Berlin et al., 1980[19]
carbon tetrachloride (5)	1.5	Baselt, 1980[15]
methylene chloride (100)	20.0	Stewart et al., 1976[46]

explains why during exposure the uptake of toluene increases with the size of adipose tissue.[18]

The same factors which increase uptake, decrease pulmonary excretion after the end of exposure. Solvents with high solubility in blood have a slow pulmonary excretion, and those with low solubility have a rapid pulmonary excretion, but as Figure 2 shows, benzene and methylene chloride, with approximately the same blood/gas partition coefficients[17] have different clearance half-lives. As a result of its longer half-life, the concentration of benzene in breath 16 hours after exposure can be used as an exposure index.[19] A similar preshift breath analysis was advocated for trichloroethylene.[16]

The use of whole organs, teeth, or hair for direct exposure tests is more limited. Harvey et al.[12] devised a noninvasive neutron activation technique for the measurement of cadmium in liver. This method may give a better estimate of body burden than other exposure tests for cadmium. Estimation of lead in deciduous teeth has been widely used in epidemiological studies because lead in deciduous teeth indicates the integrated exposure between the eruption of a tooth and its removal.[20-22]

Hair is a reasonably good index medium for metals when external contamination, if present, can be removed. Thus, hair lead has been used as an index of exposure both in industrial workers[23] and in the general population, including children.[24] Other metals which can be estimated in hair are Hg, Cd, Ni, Cu, and Mn,[25] but hair is the most important test medium for the estimation of methylmercury exposure. Mercury concentration in hair at the time of its emergence from the scalp is approximately 250 times higher than in blood, and, therefore, longitudinal analysis in 1-cm segments gives a printout of monthly blood concentrations.[26] Figure 3 shows segmental analysis on two hair samples which are representative of two different types of exposure. The upper part of the figure shows the segmental analysis of hair from a victim of the Iraqi methylmercury epidemics.[26] The temporary exposure was high and ended abruptly when the government confiscated methylmercury-dressed wheat stocks. The decline from the maximum concentration toward the scalp indicates approximately 70 days half-life. The next sample is from a North American Indian girl. The seasonal variation of methylmercury in her hair followed the yearly fishing season.

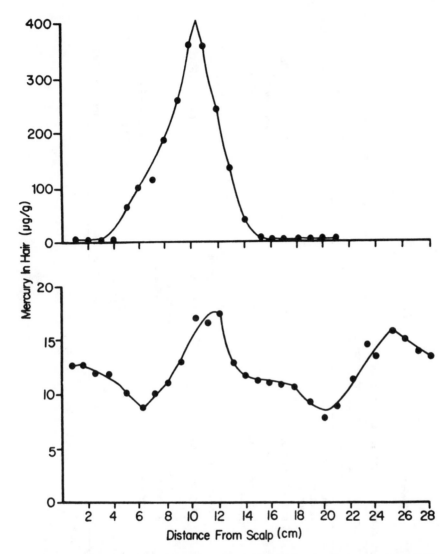

Figure 3.   Segmental hair analysis of mercury as an indicator of methylmercury exposure. The upper curve shows the analysis of hair taken from a victim of the Iraqi methylmercury epidemics.[26] The lower curve shows the mercury concentration in hair obtained from a North American Indian girl.

Indirect and Clinical Exposure Tests

Table 6 lists indirect and clinical laboratory tests. Some of them, like ALAD, cholinesterase, and methemoglobin, can indicate the degree of exposure below the toxic exposure range and even below the TLV. Others,

Table 6. Indirect and Clinical (Laboratory) Exposure Tests in Blood and Urine

Substance	Indicators		Reference
	In Blood	In Urine	
aromatic amino and nitro compounds	methemoglobin verdoglobin Heinz bodies		Pacseri and Magos, 1958[4]
cadmium	acid phosphatase	β-microglobulin	Buchet et al., 1980[27]
carbamates	cholinesterase		Baselt, 1980[15]
dibromochloropropane	follicle stimulating hormone		Whorton et al., 1979[28]
lead	ALAD zinc-protoporphyrin	ALAU coproporphyrin	Zielhuis, 1974; 1978[1,3]
mutagens	cytogenic analysis in cultured lymphyocytes, alkylated hemoglobin	mutagenicity, alkylated amino acids	Vainio et al., 1981[31]
organophosphorus	cholinesterase		Baselt, 1980[15]
PCB		copro- and uroporphyrins	Colombi et al., 1982[48]
PVC, polyurethanes		5-hydroxyindolic acid	Koricana et al., 1982[49]

like β-microglobulinuria indicates renal dysfunction[27] and an increase in the serum level of follicle stimulating hormone is the predictor of azoospermia.[28] Cytogenic analysis of peripheral lymphocytes is a routinely acceptable method for the indication of genetic damage caused to somatic cells by a wide range of mutagens, but results must always be evaluated in relation to matched controls.[29-30] The main confounding variable for chromosome aberrations is radiation. For the mutagenicity of urine and, to a lesser extent, for sister chromatid exchanges,[31] the confounding factor is smoking. Alkylation of hemoglobin is a very sensitive index of exposure to compounds which alkylate (e.g., ethylene oxide[32]) or which in association with another compound (e.g., nitrite with aminopyrine) forms an alkylating agent.[33] The combination of gas chromatography and mass spectrometry can estimate 10^{-11} g alkylated amino acids.

Other Diagnostic Methods in the Assessment of Exposure

Besides clinical laboratory methods, nonlaboratory diagnostic techniques can also be used in the assessment of exposure. The radiological demonstration of lead-line in the long bone of children[34] or decrease in the nerve conduction velocity of lead workers[35] were observed below 0.7 mg/L blood lead, previously believed to be safe. Assessment of the effects of toxic compounds on fertility using sperm counts[28] or questionnaires[36] also belong to this category. Because of lack of specificity, toxic effects often cannot be evaluated without a matched control population. Examples are studies aimed to reassess the borderline between toxic and nontoxic exposure to methylmercury. In these studies, paresthesia and other nonspecific neurological signs are correlated with the hair and blood concentrations of mercury. In a Peruvian population chronically exposed to methylmercury from the long-term consumption of marine fish, the frequency of paresthesia was as high as 29.5%. However, the role of methylmercury as a responsible agent was excluded because 49.5% of the control population of a neighboring village with a five times lower fish consumption had paresthesia.[37]

WASTE DISPOSAL SITES, EXPOSURE, AND EPIDEMIOLOGY

The migration of toxic substances from a waste dump produces three-dimensional concentration isopleths for each substance. The shape of isopleths depends on the physical and chemical characteristics of the substance and on geological-hydrological conditions. (See Chapters 2 and 3.) Volatilization, erosion of surface soil, and migration to community water sources are mainly responsible for bringing a dumped substance into contact with the communities. While the toxic hazard, the group of workers exposed to a toxic agent, and even the possible form of contact are usually known in the case of occupational exposure, these data must be collected before the assessment of exposure by biological monitoring of chemicals emitted from waste sites. Thus environmental and biological monitoring are not alternatives. The former can establish the need for the latter, can limit the number of substances to be monitored, and can also limit the size of the test population. Biological monitoring carried out within these limits helps to avoid unnecessary inconvenience to groups which could not be at risk and minimize the inconvenience to those who may be at risk.

As nearly all biological monitoring methods have been developed for the evaluation of industrial exposure,[15,38] they must be selected and used judiciously. The concentration of many substances (mainly solvents) declines in biological media so rapidly that samples must be collected from industrial workers at the end of a shift. When collection time can be extended, estimated values are extrapolated to last exposure. When the substance has a long clearance half-life, instead of the time of sampling, the background concentration (exposure) is the critical problem in the evaluation of exposure. The estimation of exposure from the dump site cannot be done without the estimation of internal exposure in comparable control groups, as the following hypothetical example shows. A real monitoring project[39] has found a significant difference in the median lead blood concentration of teachers in Tokyo (60 μg/L) and in Mexico City (220 μg/L). If the hypothetical epidemiological study finds 220 μg Pb/L blood in the study populations of both cities and the control values are the same as in teachers, the contribution of dump site to total lead exposure is significant in Tokyo and nil in Mexico City.

The relationship of exposure parameters to dose (exposure)-dependent response is used to predict the consequences of exposure to health. Any prediction is based on the assumption that exposure is correctly estimated and exposure below the biological threshold limit for a specific response assures protection. It may be that neither of these assumptions is correct. The first assumption may be wrong because (1) exposure estimated at a certain point of time is not representative of exposure for a longer period, (2) exposure to a crucial substance was not identified, and (3) sampling and analytical errors distorted results. The second assumption may be wrong because (1) the threshold limit was based on insufficient data, (2) multiple exposures altered the biological threshold limit for the substance, and (3) the biological threshold limit developed for industrial workers is not valid for the group under investigation. First of all, the pattern of industrial exposure is different from the exposure to toxic agents present in and around the home. A community exposed to toxic agents has old people, children, and housewives who may have longer exposure times and shorter exposure-free periods than usual in industry. Moreover, they may be more sensitive to the toxic agent than industrial workers. However, the need for an epidemiological study to investigate the possible effects of exposure may arrive even when neither the validity of exposure assessment nor the validity of threshold limits is in question because (1) the exposure is too high or at least too near to the toxic limit of exposure, or (2) there is a need to reassure the population about its health.

The migration of a toxic substance from a dump site, the turnover of the population, and differences in their daily patterns of life assure that neither the degree of exposure nor exposure time is homogeneous. This presents a problem for the selection and grouping of the study population. The advantage for statistics offered by extending the size of population and the disadvantage presented by stretching the borderline too far for one group (or for the whole study population) must be balanced. When exposure is above the limit of toxic threshold and exposure produces a specific response, the population size can be smaller than when the response is nonspecific. Thus the extreme rarity of hepatic angiosarcoma helped to establish the association between exposure to vinyl chloride and this disease. When the type of cancer is not a rarity, statistical uncertainties are so great that even in animal experiments a cancer risk below 10% cannot be accurately characterized.[40] Besides statistical difficulties, the length of the latent period is also an important problem when a study aims to investigate a long-term effect such as cancer. That is why screening methods, e.g., the alkylation of Hb or cytogenic analysis, for the detection of responses to mutagens with a potential to cause cancer are so important.

Most of the responses to toxic exposures become manifest within a more manageable time than cancer. The followup for the detection of fetal and child developmental defects requires a few years. The importance of such a study is that toxic agents which pass through the placenta may damage the fetus even when the mother remains asymptomatic, as has been shown[41] for methylmercury. At the extreme end of the detectability time scale are those agents which immediately produce a response, like respiratory irritation by SO_2 and dermal irritation by chlorinated hydrocarbons.

Irrespective of the response or biological test, the reliability of conclusions drawn from the results depends on the validity of the study. The observation of guidelines to improve validity aspects of studies in toxicological epidemiology[42-43] helps to increase the chances of success.

CONCLUSION

Biological monitoring is one form of internal exposure assessment. In biological monitoring, either the concentration of the toxic substance or metabolites is measured or exposure is defined by an established correlation between some biochemical effects (e.g., methemoglobin formation, cholinesterase inhibition) and exposure. Only those indirect tests which can assess exposure around the TLV and below the toxic exposure range are used for biological monitoring. However, it is often necessary to assess toxic exposure or identify exposure by its subclinical and clinical effects. Though some indirect exposure tests cover not only subtoxic but overtly toxic exposure ranges (e.g., COHb), it may be necessary to include the use of other clinical diagnostic tests in the arsenal of methods. The need to use tests which assess exposure through toxic effects mainly arises when (1) exposure is unknown or complex, but the responsible agent can be identified by diagnostic methods, (2) when exposure during a long-term period was not measured, and direct and indirect exposure tests cannot give information on past exposure, and (3) when the TLV requires reassessment.

REFERENCES

1. Zielhuis, R. L. Biological Monitoring. *Scand. J. Work Environ. Health* 4: 1–8, 1978.
2. Elkins, H. B. The Chemistry of Industrial Toxicology. Second ed., John Wiley & Sons, New York, 1959.
3. Zielhuis, R.L. Biological Quality Guide for Inorganic Lead. *Int. Arch. Arbeitsmed* 32: 103–127, 1974.
4. Pacseri, I. and Magos, L. Determination of the Measure of Exposure to Aromatic Nitro and Amino Compounds. *J. Hyg. Epidemiol. Microbiol. Immunol.* 2: 99–110, 1958.
5. Magos, L. Haemoglobin Analysis of TNT Workers. *Proc. 14th Intern. Congr. Occup. Health* 2: 838–840, Madrid, 1963.
6. Jones, C. T. A. and MacKay, H. F. A. Carbon Monoxide Poisoning in a Former Mining Community. *Brit. Med. J.* 286: 603–604, 1983.
7. Harvey, T. C., Chettle, D. R., Fremlin, J. H., AlHaddad, I. K., and Downey, S. P. M. J. Cadmium in Shipham. *Lancet* 1: 896-899, 1979.
8. Philipp, R. and Hughes, A. O. Health Effects of Cadmium. *Brit. Med. J.* 282: 2054, 1981.
9. Inskip, H., Beral, V. and McDowall, M. Mortality of Shipham Residents: 40 Years Follow Up. *Lancet* 1: 896–899, 1982.
10. Maltoni, C. Occupational Carcinogenesis. Cancer Detection and Prevention. International Congress Series No. 322, 19–26, Excerpta Medica, Amsterdam, 1973.
11. Creech, J. L. and Johnson, M. N. Angiosarcoma of Liver in the Manufacture of Polyvinyl Chloride. *J. Occup. Med.* 16: 150–151, 1974.
12. Harvey, T. C., McLean, J. S., Thomas, N. J. and Fremlin, J. H. Measurement of Liver Cadmium in Patients and Industrial Workers by Neutron-Activation Analysis. *Lancet* 1: 1269–1272, 1975.
13. Wigaeus, E., Holm, S. and Åstrand, I. Esposure to Acetone. Uptake and Elimination in Man. *Scand. J. Work Environ. Health* 4: 84–94, 1981.
14. Åstrand, I., Övrum, P., and Carlsson, A. Exposure to Methylene Chloride. I. Its Concentration in Alveolar Air and Blood During Rest and Exercise and its Metabolism. *Scand. J. Work Environ. Health* 1: 78–94, 1975.
15. Baselt, R. C. Biological Monitoring Methods for Industrial Chemicals. *Biomed. Publ.*, Davis, CA, 1980.

16. Lauwerys, R. Biological Criteria for Selected Industrial Toxic Chemicals. *Scand. J. Work Environ. Health* 1: 139–172, 1975.
17. Åstrand, I. Uptake of Solvents in the Blood and Tissues of Man. *Scand. J. Work Environ. Health* 1: 199–218, 1975.
18. Carlsson, A. and Lindquist, T. Exposure of Animals and Man to Toluene. *Scand. J. Work Environ. Health* 2: 199–211, 1976.
19. Berlin, M., Gage, J. C., Gullberg, B., Holm, S., Knutsen, P. and Tunek, A. Breath Concentration as an Index of Health Risk from Benzene. Studies on the Accumulation and Clearance of Inhaled Benzene. *Scand. J. Work Environ. Health* 6: 104–111, 1980.
20. Altshuller, L. F., Halak, D. B., Landing, B. H. and Kehoe, R. A. Deciduous Teeth as an Index of Body Burden of Lead and Radium. *J. Pediatr.* 60: 224–229, 1962.
21. Fosse, G. and Berg Justesen, N.-P. Lead in Deciduous Teeth of Norwegian Children. *Arch. Environ. Health* 33: 166-175, 1978.
22. Needleman, H. L., Tuncay, O. C., and Shapiro, I. M. Lead Levels in Deciduous Teeth of Urban and Suburban American Children. *Nature* 235: 11–112, 1972.
23. Baloh, R. W. Laboratory Diagnosis of Increased Lead Absorption. *Arch. Environ. Health* 28: 198–208, 1974.
24. Phil, R. O. and Parker, M. Hair Element Content in Learning Disabled Children. *Science* 198: 204–206, 1977.
25. Eads, E. A. and Lambdin, C. E. A Survey of Trace Metals in Human Hair. *Environ. Res.* 6: 247–252, 1973.
26. Clarkson, T. W., Amin-Zaki, L. and Al-Tikriti, S. An Outbreak of Methylmercury Poisoning Due to Consumption of Contaminated Grain. *Fed. Proc.* 35: 2395–2399, 1976.
27. Buchet, J. P., Roels, H., Bernard, A. Jr., Lauwerys, R. Assessment of Renal Function of Workers Exposued to Inorganic Lead, Cadmium and Mercury Vapor. *J. Occup. Med.* 22: 741–750, 1980.
28. Whorton, D., Milby, T. H., Krauss, R. M. and Stubbs, H.A. Testicular Function in DBCP Exposed Pesticide Workers. *J. Occup. Med.* 21: 161–166, 1979.
29. Sram, R. J. and Kuleshov, N. P. Monitoring the Occupational Exposure to Mutagens by the Cytogenic Analysis of Human Peripheral Lymphocytes In Vivo. *Arch. Toxicol.* Suppl. 4: 11–18, 1980.
30. Sram, R. J., Zudova, Z. and Kuleshov, N. P. Cytogenic Analysis of Peripheral Lymphocytes in Workers Occupationally Exposed to Epichlorohydrin. *Mutation Res.* 70: 115–120, 1980.
31. Vainio, H., Sorsa, M., Rantanen, J., Hemminki, K. and Aitio, A. Biological Monitoring in the Identification of the Cancer Risk of Individuals Exposed to Chemical Carcinogens. *Scand. J. Work Environ. Health* 7: 241–251, 1981.
32. Calleman, C. J., Ehrenberg, L., Osterman-Golkar, S., Segerblack, D., Svenson, K., and Wachtmeister, C. A. Monitoring and Risk Assessment by Means of Alkyl Groups in Hemoglobin in Persons Occupationally Exposed to Ethylene Oxide. *J. Environ. Pathol. Toxicol.* 2: 427–442, 1978.

33. Farmer, P. B. Monitoring for Human Exposure to Carcinogens. *Chemistry in Britain* 18: 790–794, 1982.
34. Betts, P. R., Astley, R., and Raine, D. N. Lead Intoxication in Children in Birmingham. *Brit. Med. J.* 1: 402-406, 1973.
35. Seppäläinen, A. M., Tola, S., Hernberg, S., and Kock, B. Subclinical Neuropathy at "Safe Levels" of Lead Exposure. *Arch. Environ. Health* 30: 180–183, 1975.
36. Levine, R., Symons, M. J., Balogh, S. A., Arndt, D. M., Kaswandik, R. N., and Gentile, J. W. A Method for Monitoring the Fertility of Workers. 1. Method and Pilot Study. *I. Occup. Med.* 22: 781–791, 1980.
37. Turner, M. D., Marsh, D. O., Smith, J. C., Inglis, J. B., Clarkson, T. W., Rubie, C. E., Chiriboga, J., and Chiriboga, C. C. Methyl Mercury in Populations Eating Large Quantities of Marine Fish. *Arch. Environ. Health* 35: 367–378, 1980.
38. Roi, R., Town, W. G., Hunter, W. G., and Alessio, L. Occupational Health Guidelines for Chemical Risk. Commission of the European Communities, Luxembourg, 1983.
39. Vahter, M. (Ed.) Assessment of Human Exposure to Lead and Cadmium Through Biological Monitoring. National Swedish Institute of Environmental Medicine and Karolinska Institute, Stockholm, 1982.
40. Peto, R. Carcinogenic Effects of Chronic Exposure to Very Low Levels of Toxic Substances. *Environ. Health Pers.* 22: 155–159, 1978.
41. Marsh, D. O., Myers, G. J., Clarkson, T. W., Amin-Zaki, L., and Tikriti, S. Fetal Methylmercury Poisoning: New Data on Clinical and Toxicological Aspects. *Trans. Amer. Neurol. Assoc.* 102: 1–3, 1977.
42. Hernberg, S. Evaluation of Epidemiologic Studies in Assessing the Long-Term Effects of Occupational Noxious Agents. *Scand. J. Work Environ. Health* 6: 163–169, 1980.
43. Zielhuis, R. L. and Verberk, M. M. Validity of Biological Tests in Epidemiological Toxicology. *Int. Arch. Arbeitsmed.* 32: 167–190, 1974.
44. Peterson, J. E. and Stewart, R. D. Predicting the Carboxyhemoglobin Levels Resulting from Carbon Monoxide Exposures. *J. Appl. Physiol.* 39: 633–636, 1975.
45. Rylander, R. and Vesterlund, J. Carbon Monoxide Criteria, with Reference to Effects on the Heart, Central Nervous System and Fetus. *Scand. J. Work Environ. Health* 7 Supp. 1: 1–39, 1981.
46. Stewart, R. D., Hake, C. L., and Wu, A. Use of Breath Analysis to Monitor Methylene Chloride Exposure. *Scand. J. Work Environ. Health* 2: 57–70, 1976.
47. World Health Organization Study Group. Recommended Health-Based Limits in Occupational Exposure to Heavy Metals. WHO, Geneva, 1980.
48. World Health Organization Task Group, Mercury. Environmental Health Criteria 1, WHO, Geneva. Health Criteria 1, WHO, Geneva, 1976.
49. Colombi, A., Maroni, M., Ferioli, A., Castoldi, M., Miu Ke Jun, Valla, C., and Foa, V. Increase in Urinary Porphyrin Excretion in

Workers Exposed to Polychlorinated Biphenyl. *J. Appl. Toxicol.* 2: 117–121, 1982.

50. Koricana, Z., Stankovic, B., Milanovic, L., Dugandzic, M. and Trkovic, T. Urinary Excretion of 5-Hydroxyindolic Acid in Occupational Exposure to Polyvinyl Chloride and Polyurethane Foams. *J. Appl. Toxicol.* 2: 213–214, 1982.

SECTION III

Determining Human Health Effects

CHAPTER 6

Health Risks of Concern

Roy E. Albert

INTRODUCTION

Virtually every chemical that is manufactured, including synthetic intermediates, ends up, to some extent, in waste dumps. Consequently, the range of potential health effects from high-level exposure to chemicals in dumps is tremendous. The nature of the chemicals in any given dump reflects the nature of the local industry, but there are also chemicals that are generally encountered in industry.

Exposure to chemicals in dumps generally involves low concentration levels. There are exceptions, such as Love Canal, where the dump was covered over by a school with a playground. When chemicals surfaced, children picked up raw chunks of lindane and phosphorus and threw them around and sloshed through liquid organic wastes. This is not a very common occurrence, and the overwhelming majority of exposures to populations around dump sites are low.

The effects that are of concern at low levels of exposure are those that are irreversible. Reversible effects reflect a balance between injury and repair processes. With decreasing dose, injury decreases to a level where repair processes are able to reverse the damage, and the deleterious health effects essentially disappear at low levels of exposure. However, with agents that

J. B. Andelman and D. W. Underhill (Editors), *Health Effects from Hazardous Waste Sites*
© 1987 Lewis Publishers, Inc., Chelsea, Michigan – Printed in U.S.A.

produce irreversible damage, namely, damage to DNA, the injury can be permanent and cumulative.

Carcinogenic effects at low levels of exposure currently dominate most of the concerns about the health effects of exposure to chemical dumps. The practical problem with carcinogens in chemical dumps is the need to develop a rational basis for setting standards to guide cleanup action and to assure exposed individuals that their health is not in jeopardy.

DEVELOPMENT OF STANDARDS

The approach to standard setting for toxic chemicals that produce chronic damage is very uncertain. There are no accepted approaches at the present time. The oldest attempts to set toxicant standards are in the field of occupational health. A number of years ago, Henry Smyth described the evolution of occupational standards.[1] During what might be called the age of chaos, every experienced industrial hygienist had a few values uniquely his own, drawn from his own experience. For less familiar substances, he borrowed more or less judiciously from values cherished by his professional colleagues. Some degree of unanimity was brought about when the U. S. Public Health Service values, based on long-term collective experience in industrial hygiene, were published in 1943. In 1947, the American Conference of Governmental Industrial Hygienists published its first list of standards in the *Industrial Hygiene* newsletter. Subsequently, they have published revised values for their standards. The 1968 values were adopted by the Occupational Safety and Health Administration (OSHA). It is interesting that no methodological basis is set for the choice of threshold limit values for these standards, even to this day.

Another route to the development of standards involved the use of ionizing radiation and radioactive materials, going back to the 1920s, with whole body radiation. Other radiation standards had to do with bone cancer induced by radioactive material such as radium and lung cancer induced by radon.

In the case of radon-induced lung cancer, a safety factor of 300 was applied to average levels of exposure in the Joachimsthal and Schneeberg uranium mines, where there was a 50% incidence of cancer. During World War II, because of the need for uranium, this originally proposed standard was raised by a factor of 30 resulting in a high incidence of lung cancer in domestic uranium miners.

The permissible body burden for radium, and by extension for other bone-seeking isotopes, was set at one-tenth the smallest amount of radium that was associated with bone cancer in one radium dial painter. The standards for whole body radiation were based initially on the dose required to produce skin erythema; the standards were progressively lowered as new information on radiation effects was obtained. These approaches, then, were a matter of defining a level of response associated with a level of exposure and applying a safety factor to define a standard.

The no-observed-effect level safety factor approach is commonly used to set standards for noncarcinogenic toxicants, but this approach has been used occasionally for carcinogens. For example, the standard for benzo[a]pyrene in the Soviet Union was formulated on the basis of an experiment in which rats were given ten monthly intratracheal intubations of benzo[a]pyrene at various levels and then observed for life for the development of tumors.[2] The highest dose that produced no tumors was used as a basis for defining an air level which, if breathed by humans for 50 years, would produce the same lung burden. A safety factor of 100 was then applied, but the resulting standard was regarded as impractically stringent, and the safety factor was reduced to 10. The standard for benzo[a]pyrene turned out to be about the average concentration of benzo[a]pyrene in the city of Kiev when this standard was developed.

There is another approach to carcinogen standard setting that grows out of risk assessment. Guidelines were formulated for the assessment of carcinogenic risks by the U. S. Environmental Protection Agency (EPA) in 1976, and these guidelines are still in place.[3] There are two aspects of risk assessment: the qualitative and the quantitative. The qualitative deals with how likely an agent is to be a human carcinogen, and the quantitative deals with how much cancer would be produced at a given level of exposure.

The qualitative aspect of risk assessment involves a weight of evidence approach based on animal and epidemiologic data that takes into account the scope of the data and the quality and nature of the responses, using short-term tests as a supportive basis for making judgments. The qualitative aspect of risk assessment is of concern because, although many dump site chemicals are generally recognized carcinogens, there are others, particularly chlorinated solvents and pesticides, which figure heavily in a number of dump sites where the qualitative evidence for carcinogenicity is disputed.

The quantitative side of risk assessment uses an assumed dose-response relationship together with exposure estimates as the basis for predicting the magnitude of risk. Because of species variability in susceptibility to carcinogens, there are large uncertainties in the use of animal data to predict human responses. A central issue is the choice of dose-response model, since we are concerned with exposures that are far below those that can be demonstrated to produce observable effects in epidemiologic studies or animal experiments. We are, therefore, forced to rely on theory as the support for the choice of extrapolation model.

The dominant extrapolation model for carcinogens is the linear nonthreshold dose-response model. It surfaced in the 1950s in connection with observations of leukemia induction in atom bomb survivors where the dose-response relationships were consistent with a linear nonthreshold dose-response model. This approach has been carried into the area of chemical carcinogens and is with us today. It provides the conceptual basis for concern with very low levels of exposure because with this model, any exposure, however small, produces excess cancer. To be sure, the lower the dose, the smaller the risk of cancer.

More recently, another rationale for the linear nonthreshold extrapolation model has been advanced, namely, that if a carcinogen has the same mechanism as whatever it is that is producing background cancers, the addition of the carcinogen causes an essentially linear increment in the occurrence of cancer. This is true even if the dose-response for carcinogenic effect is nonlinear; for practical purposes, within one or two multiples of the background level, the dose-response will be essentially linear.

The outcome of quantitative risk assessment is twofold. One is an estimate of personal risk, that is, the excess risk to the individual given a level and duration of carcinogen exposure, as well as the risk to the exposed population in terms of excess numbers of cancer cases.

The use of quantitative risk assessment is a source of contention, and it has its very vigorous opponents. The same groups can sometimes be opponents and sometimes proponents of quantitative risk assessment. At one time, OSHA was opposed to the use of quantitative risk assessment, arguing that it was not necessary for regulations that are based on technical and economic feasibility. This position was not supported by the Supreme Court and, more recently, OSHA has become intensively involved with quantitative risk assessment. Industry has opposed quantitative risk assessment on occasion and uses it on other occasions, particularly when the risks are very low. Individuals who use quantitative risk assessment appreciate that it generally has meaning only at the extremes. When the risk is exceedingly low, quantitative risk estimation gives reassurance, since the use of the linear nonthreshold extrapolation model tends to be conservative in providing a plausible upper limit for risk estimates, i.e., the calculated risk is not likely to be greater, and it could be less than the actual risk. At the other extreme, if the risks turn out to be very high, the estimates prompt consideration of further efforts to reduce exposure.

Quantitative assessment has different kinds of uses. It has figured into the weighing of risks and benefits. It has been important in priority setting. It has been useful in terms of evaluating the residual risk after the application of best available technology. The use of quantiative risk assessment is a promising basis for the setting of standards to irreversibly acting agents such as carcinogens. For example, in the EPA, there is no legislative basis for using economic or technical considerations in setting water quality criteria. The approach taken for carcinogens was to recommend levels that were associated with risks in the order of 10^{-5} to 10^{-7} lifetime excess cancer risk. The U. S. Food and Drug Administration (FDA) has proposed a standard for residual carcinogens in the edible portions of food animals on the basis of a lifetime excess cancer risk level of 10^{-6}.

Carcinogen standards for the general public ought to be in reasonable balance with occupational carcinogen standards. OSHA has set carcinogen standards for vinyl chloride, acrylonitrile, benzene, arsenic, and coke oven emissions. The 30-year occupational lifetime cancer risks associated with these standards can be calculated using the EPA's Carcinogen Assessment Group's potency values based on the linear nonthreshold dose-response model. At the current OSHA standards, the lifetime risk for acrylonitrile is 2.5%, benzene is 1.5%, coke oven emission is 1.3%, whole body ionizing radiation is 1.5%, radon is about 4.3%, vinyl chloride is 0.07%, and arsenic is 0.2%.[4]

These standards were set without reference to cancer risks. They were were set on the basis of technical and economic feasibility. In the atomic energy industry, the standard of 5 rems per year is actually not followed in practice since i t has been feasible to achieve an average exposure about 0.5 rem per year which is associated with a lifetime cancer risk of 0.15%.

It would probably be feasible to develop a general standard for occupational carcinogen exposure at a risk level of 0.1%. It seems unbalanced to control general population exposures at a risk of 0.0001% (10^{-6}) when industrial workers are exposed to carcinogen risks as high as 2 to 4%. A more balanced approach between the occupational and public carcinogen standards would involve a 10^{-3} lifetime cancer risk for industrial workers for individual carcinogens and a 10^{-5} lifetime cancer risk for individually regulated carcinogens for the general public.

CONCLUSION

Armed with a set of standards, the problem of dealing with chemical dumps would be simplified, because it is particularly difficult to deal with low-level exposures without any acceptance of what constitutes a permissible limit of exposure. The problem with setting carcinogen standards based on risk, is the concept that there is no such thing as a safe exposure to carcinogens. We are, therefore, forced into defining what is an inconsequential risk.

When one considers the notion of what is an inconsequential risk, one thinks of a drop in a bucket. How much is a drop in a bucket? Buckets contain about 10 liters and a drop is about .06 mL, so the ratio of a drop to the contents of a bucket is about 6×10^{-6}. Hence, a risk level of 10^{-5} (the suggested carcinogen standard for the general public) seems fairly inconsequential since it amounts to no more than two drops in a bucket.

REFERENCES

1. Smyth, H.F. Jr. Improved Communication: Hygienic Standards for Daily Inhalation. *Am. Ind. Hyg. Assoc.* Quarterly 17:129–185, 1956.

2. Yanysheva, N.Ya. and Antomonov, Yu.G. Predictign the Risk of Tumor Occurrence Under the Effect of Small Doses of Carcinogens. *Environ. Health Pers.* 13:95–99, 1976.
3. Albert, R.E., Train, R.E., and Anderson, E. Rationale Developed by the Environmental Protection Agency for the Assessment of Carcinogenic Risks. *J. Nat. Cancer Inst.* 58(5):1537–1541, 1977.
4. Albert, R.E. The Acceptability of Using the Cancer Risk Estimates Associated with the Radiation Protection Standard of 5 Rems/Year as the Basis for Setting Protection Standards for Chemical Carcinogens with Special Reference to Vinyl Chloride. Prepared for the Ministry of Labour, Occupational Health and Safety Division, Toronto, Ontario, Canada, 1982.

CHAPTER 7

Expectation and Limitation of Human Studies and Risk Assessment

Marvin A. Schneiderman

INTRODUCTION

The U. S. Supreme Court, in its 1979–1980 term, was active in areas of environmental law; the best known decision in the health regulation area was that handed down in July 1980 in *AFL-CIO* v. *American Petroleum Institute*, the so-called "Benzene Case." The Court ruled, in a plurality opinion written by Justice Stevens, that the Occupational Safety and Health Administration (OSHA) must make the finding that regulated substances pose a "significant risk" of material health impairment. The Court did not specify what data would, in the minds of the three justices who joined Justice Stevens in his entire opinion, constitute evidence of significant risk. (Mr. Justice Powell joined Mr. Stevens, to make a majority, but he wrote a separate opinion.) Nor did the Court resolve the issue of cost-benefit, noting only that the fact that a material was shown to be a carcinogen did not necessarily mean that any level of exposure to that material posed a significant risk. The Court did require that OSHA quantify risks, but only Mr. Powell supported the finding of the Fifth Circuit Court of Appeals that benefits of regulation be "reasonably related" to the costs of regulation.

J. B. Andelman and D. W. Underhill (Editors), *Health Effects from Hazardous Waste Sites*
© 1987 Lewis Publishers, Inc., Chelsea, Michigan – Printed in U.S.A.

OSHA, in reevaluating its earlier Carcinogen Policy[1] gave among the reasons for reexamination, "to assure consistency with Supreme Court decisions" and also "to consider and respond to any advances or changes in the science of carcinogenesis, including quantitative risk assessment, occurring since the closing of the record in 1979."

In a compilation of comments to OSHA's invitation[2] almost forty pages are devoted to cost-benefit (although the Court did not require this, Executive Order 12291 apparently did) and to techniques of quantitative risk assessment and significant risk determinations. (A summary of these comments is given in Appendix D.)

Agencies other than OSHA are also concerned with issues of risk assessment. The U.S. Food and Drug Administration (FDA), "in response to a directive from the Congress of the United States . . . contracted with the National Academy of Sciences to conduct a study of the institutional means for risk assessment." One of the major questions asked of the NAS, and one also addressed in the comments to OSHA, was "should the federal government establish a separate agency for the specific (and only) purpose of conducting risk assessments?" EPA has informally (and operationally) modified the policy contained in the Interagency Regulatory Liaison Group (IRLG) statement.[3] The Office of Science Technology Policy has circulated for comment a draft of the first part (the "science base") of what is intended to be a revised federal government cancer policy. Several states, most notably California, have been making attempts at their own cancer policies. However, only where there are specific references to risk assessments are these policies considered here.

An attempt is made here to apply the NAS/NRC recommendations to a single (complex) material, i.e., asbestos, for which a substantial amount of human data exists. This is followed by a consideration of what might be a simpler problem, if a simpler material, i.e., formaldehyde, were considered. The last part of this chapter considers the general issue of the potential availability of data.

NAS/NRC RECOMMENDATIONS

In the March 1, 1983 report, the NAS/NRC committee saw the risk assessment process as falling into four successive steps:

1. Hazard identification: Is the substance health damaging?
2. Dose-response assessment: How much of the substance causes how much damage?
3. Exposure assessment: How many people exposed to how much of the substance, and if it is regulated, how many fewer people will then be exposed?

4. Risk characterization: After the proper manipulations of items 2 and 3 are made, what is the resulting magnitude of human risk? What is the uncertainty attached to this estimate of magnitude?

The NAS/NRC committee placed most of its emphasis on cancer, not because cancer is the most important disease, or the most expensive, or causes most deaths, or because cancer patients occupy the most beds, or even because cancer is the most feared of diseases. They considered cancer because there was the most scientific information on cancer. In addition to all the vital statistics data, there are relatively good animal models of the disease; there are workable theories about the development of the disease (such as the multistage theory) that are consistent with much of the data; there are mathematical models of dose-response relationships that permit some risk assessment; and there is a vast literature full of enough shades of opinion and variety of ideas to provide a researcher or an administrator with a first-class paper refuge (and space to pray that the wolf doesn't come and blow it all down). There is information on hundreds of materials that have been tested by a score or more of different tests, so that we can sometimes say "material X is (or is not) likely a carcinogen" in some animal species, and sometimes in humans.

There are also questions, of course. In an animal test system, is the animal representative for man? (Perfectly safe answers usually come in pairs to these questions: Answer 1 – "I don't really know." Answer 2 – "Possibly, but not under all circumstances. A human is not a featherless hen, you know." Of course, I know. What made the answerers think that anyone ever thought humans were featherless hens?) Should "benign" tumors be counted? Is the mouse liver tumor a real cancer? How about adenomas of the mouse lung? Injection site sarcomas? Do you accept the positive result of a single experiment at one dose in one sex in one strain or species of animal? (Of course, it's there whether I accept it or not.) How many negative tests does it take to outweigh one positive test? If the tests have the same power, and the significance test conducted was at the 0.05 level, I would say 19 plus 1. If I were Dr. Pangloss and believed in a perfect world (no false positives and no false negatives), I might say one or two. Are materials that do not cause mutations in the short-term tests really carcinogenic? Shouldn't I have at least two positive short-term tests before I call a material a mutagen, or genotoxic? And how many negative tests should I have on a material before I could feel safe if my wife, child, mother-in-law, or pet kinkajou should be exposed to it?

How about epidemiologic data? Must I have a relative risk of 2 or more before I am satisfied that I have something? For both sexes? For all races? Must I know all about cigarette smoking (including tar content of the cigarettes smoked) before I can say anything about lung cancer? Bladder cancer? Pancreatic cancer? Kidney cancer? Shouldn't I know something

bestos exposure, too? Do I have to have data on some of the
ility enzyme systems like AHH? If I don't have all this will I
nonetheless have enough information to make a (scientific) recommendation to an administrator, or a regulator—other than "Wait. We need more
data." (And add, prayerfully, "And, please, no one get sick or die from this
stuff until I get all the data I want. At least, no one I know.") How do I
make a recommendation about materials for which there is exposure—but
not likely to be any fruitful epidemiology data?[4]

And so, having answered all these questions to my satisfaction (and
coming away with a clear conscience, so that when the two guys in the white
suits, socks, and shoes from that insurance company television commercial
come to visit me, after I've died on my feet from some nice clean, simple,
nondebilitating, painless disease, I'll be able to walk through walls and on
air with them—and with equanimity) and having made the determination
that a material, or a process, or a complex of materials, is carcinogenic or
likely to be carcinogenic for humans, I will need only to develop a dose-
response relationship[5] and then some exposure estimates.

Here I consider only two materials—asbestos, a catchall term, which
covers many varieties, sizes, chemical compositions, etc., requiring sophisticated equipment and professionals to identify it, and formaldehyde, a simple chemically identifiable material, rather abundant, and a normal resident
(with possibly a short residence time) of the human body and lots of commercial materials.

DOSE-RESPONSE—ASBESTOS

In 1981, using the work of W. Nicholson,[6] some colleagues and I reviewed
several studies which appeared to give some dose and response information
about asbestos in humans—all following industrial exposure.[7] Nicholson
had done exposure estimates, and we were able to construct estimates of the
increase in illness following a unit increase in dose. Of nine studies that at
first seemed suitable, we discarded two because of methodological problems. Two other studies of workers (exposed during mining and milling
operations) yielded much lower estimates, so these two were set aside as
probably not representative of either industrial exposure or general ambient
exposure. This left five studies (Table 1) capable of giving useful dose-
response information. As a result of eliminating four studies, the range of
response has been limited to 30-fold. (If all nine studies had been included,
the ratio of highest to lowest dose-response slope would have been 150-
fold.)

Some of the wide range of slopes may be due to difficulties in measurement. (Only one study among the nine had repeated industrial hygiene-
supervised measurements, and those were revised during the course of rean-
alysis of the data by the original author. (This study is not among the five

listed in Table 1.) Fiber size was undoubtedly different in the different studies, but little or no information is available on that. (Long fine fibers appear to be more carcinogenic than short stubby fibers.)

For lung cancer there appears to be no serious problem of what form of dose-response relationship to use; almost all research workers agree that the relationship is linear with cumulated dose, that it is independent of age (or nearly so), and that asbestos has the effect of multiplying the background cancer rate by some factor depending on dose. (For the Seidman study[8] included in Table 1, the multiplier was about 5 for smokers and nonsmokers alike). Asbestos thus appears to behave with respect to lung cancer as if it were a late-stage carcinogen. Because some asbestos bodies are generally retained in the lung (and other tissues), arguments about whether there may be a threshold—i.e., some perfectly safe exposure—are not raised, even by most people who wish to see more relaxed rather than more rigid standards for exposure. To summarize the dose-response data relating asbestos to lung cancer:

1. Slopes of dose-response curves can be derived from exposure data on humans (mostly adult males).
2. Among studies with some dose data, and with the fewest methodological problems, and which represent the major industrial (manufacturing) activities in the United States, the ratio of the highest to the lowest slope is about 30 to 1.
3. Fiber dimensions undoubtedly have an effect on dose response. Generally little or no information is available on fiber size. (Long thin fibers appear to be more effective than short stubby ones.) There are many more short fibers in the ambient atmosphere.
4. Fiber type appears to be unimportant with respect to lung cancer—but this is not certain. Most industrial exposures in the past have been mixed, i.e., to several fiber types.
5. Dose-response data are mainly for men. In a study that considered both men and women, the slope for women was 6.5 times as great as the slope for men. It is not known if women are more susceptible or whether this result is an artifact.

This summary has dealt only with lung cancer. Mesothelioma appears to act at an earlier stage in the cancer process, and its mode of action has many other characteristics different from how asbestos leads to lung cancer. If asbestos appears to act as an early-stage material (genotoxic?) with respect to mesothelioma, then it appears that the tissue–material interaction, rather than some inherent genotoxic or nongenotoxic quality of the material, is the overriding factor.

Table 1. Estimated Increases in Lung Cancer Per Unit of Exposure to Asbestos (from Nicholson[6])

Study No.	Group	Type of Asbestos	Duration From Onset	Of Exposure	Estimated Asbestos Concentration	Relative Risks (range)	Slope Percentage Increase Per f-yr/mL	Source and Notes
1	Insulation Manufacturing Patterson, NJ	Amosite	35 years	Short 1 month-2 years	35 f/mL (no direct measures)	2.5-6.5	9.1[a] 1979	Seidman et al.[8]
2	Manufacturing London, U.K.	Crocidolite Chrysotile Amosite	11-42 years (men) 33-40 years (women)	Not given	5-20 f/mL (basis not stated)	1.3-21.3	1.3 (men)[a] 8.4 (women)	Newhouse and Berry[9]
3	Manufacturing several U.S. factories (retirees)	Amosite Chrysotile some Crocidolite	9-51 years	3-51 years	Dust measure for job: conversion 1 mppcf = 3 f/mL	2.0-7.8	0.3[a]	Henderson and Enterline[10]
4	Manufacturing Manville, NJ	Chrysotile some Amosite Crocidolite	38 + years	20-38 years	10-80 f/mL company estimates for highest exposures	4.0 (lowest exposure category)	1.1% (based on lowest group)	Nicholson et al.[6] Problems of saturation (wasted dose) at highest levels)
5	Textile Manufacturing, U.S.	Chrysotile	10-36 years	>6 months	5-80 f/mL	2.2-15.5	5.3%[a]	Dement et al.[11] (High background lung cancer rates)

[a]Regression estimate.

EXPOSURE ASSESSMENT—ASBESTOS

Having gone through the first two steps of the NAS/NRC process and having arrived, rather shakily, at an estimate of dose-response (or a range of estimates), we must now ask ourselves the "how much" questions: How much exposure? To whom? When? Where? The permissible level in industrial exposures to asbestos has been dropped from 12 f/mL in 1968 to a current level of 2 f/mL—with recommendations for reductions to 0.5 f/mL. In Great Britain, the current allowable level is 1 f/mL, with some fiber types prohibited. These changing levels and standards raise problems of dose in the estimated process. Breslow et al.[12] have recently written about how to try to handle the problem of changing dose in estimating risks.

If we are concerned with ambient exposures, we need to be concerned with air, food, and water contamination, as well. A document of the Organization for Economic Cooperation and Development (OECD), dated November 16, 1982,[13] lists concerns with asbestos exposures from mining and milling, a whole series of manufacturing and end use activities, demolition and disposal of waste asbestos, and the several natural sources. OECD estimates the daily exposure to persons in an industrial society as follows (Table 2).

Table 2. Estimated Daily Exposure in an Industrial Society

	Inhaled[a] μg/day	Ingested μg/day	Total μg/day
Ambient air	0.001–0.05	0.011–0.07	0.012–0.12
Paraoccupational	0.019–1.70	0.027–1.72	0.044–3.42
Occupational	1.67–8.37	1.68–8.39	3.35–16.76

[a]1,000 fibers/ng (10^6 fibers/μg).

When one wishes to convert these levels into risks for humans, one then has to deal with the double set of uncertainties: (the 30- to 150-fold variation in dose-response curves and the 5 (occupational) to 10 (ambient air) to 75 (paraoccupational) range of variation in exposure. It is not surprising that OECD concludes that the health effects of low-level exposure "remain ambiguous."

From the data my colleagues and I developed on behalf of the Bundesgesundheitsamt (BGA) of the West German government, OECD made the estimates (Step 4 in the NAS/NRC process) of excessive lifetime risk given in Table 3.

Thus, when we try to go to NAS/NRC's Step 4, risk characterizations, we are in trouble. With the best of human data, the range of estimates of new cases of a disease (mesothelioma) largely related to *only* this exposure, the best (smallest) range of estimates is about ninefold. Of course, asbestos is not a single material, and size as well as type of asbestos is of consequence, and, if we consider the different cancers that appear to be associated with

Table 3. Asbestos Risk Assessment OECD Estimates

| | | Risks (Deaths/100,000) | | Estimated Increased Risk from Exposure (Deaths/100,000) | | |
	Disease	Background	From 1 f yr/mL Beginning at Birth	Ambient Air	Paraoccupation	Occupation
Male						
Smokers	Lung cancer	10,000	100–1,000	1.5–150	37–3,700	2,100–12,000
Nonsmokers	Lung cancer	1,000	10–100	0.2–15	3.7–370	210–1,200
Female						
Smokers	Lung cancer	4,000	40–400	0.6–60	15–1,500	850–4,000
Nonsmokers	Lung cancer	500	5–50	0.1–8	1.9–190	110–610
All	Mesothelioma	0[a]	80–480	0.7–70	17–1,700	410–3,300

[a]This is obviously an error. There are some small number of mesotheliomas that are not associated with asbestos exposure.

asbestos exposure, we also have to consider that different mechanisms of action are likely to be involved. Changing exposure may have an earlier effect on the asbestos-related lung cancer deaths than on the mesotheliomas (because of the possible late-stage effects in lung cancer). All of these factors contribute uncertainties when we try to use these data to predict, or estimate, what might happen in the future.

It is hardly any wonder that recent estimates of future cases of mesothelioma from earlier exposures (which almost all authors try to relate to excess cases of lung cancer) that may lead to lawsuits range from about 500[14] to over 3,000[15] a year.

The problems in estimating the hazard of exposure to asbestos are summarized as follows:

1. Data are for industrial populations.
2. Little information on women; one study shows 6.5 times greater response in women than men.
3. "Asbestos" is a catch-all; some types may be more hazardous than others.
4. Fiber dimensions are important:

   • Big (long, thin) fibers are more carcinogenic
   • Little (short, stubby) fibers are less toxic
   • Few data of any kind exist on specific fiber size associated with specific exposures

5. Asbestos appears to:

   • Multiply background risks
   • Operate as a late stage carcinogen — with persistent fiber retention

ANOTHER, SIMPLER MATERIAL—FORMALDEHYDE[a]

The complexities of estimating effects of exposure to asbestos arise, in part, from the complexity of the material, and the complexity and shifting nature of the exposure. Therefore it seems worthwhile to look at a far less complex material, formaldehyde, and see what we can make of it. This is also particularly timely because the U.S. Court of Appeals, Fifth District Circuit Court, in mid-April 1983 overturned the Consumer Product Safety

---

[a]Another possible material to examine might be benzene — particularly in view of the fact that the Supreme Court essentially ordered OSHA to do a benzene risk assessment. EPA, OSHA, IARC, and at least one private consultant have made (published) attempts at estimating the consequences of ambient and/or industrial exposures. Although benzene is a simpler material than asbestos, the attempts at risk assessment seem at least equally compounded with difficulty. The IARC attempt was buffeted by a "suggestion" by a federal official that the results were not "science" and should not be published. Enough was published so that anyone familiar with the decimal system can make his (her) own estimates.[16] It is also true that epidemiologic studies of benzene with good indications of exposure are rarer than the critics of those studies. Until very recently there were no reports of laboratory animal exposures to benzene leading to

Commission's restriction on the use of ureafoam formaldehyde for home and school insulation. Part of the Court's opinion cited the ubiquity of formaldehyde. "It is present in every cell in the human body and in the atmosphere." Apparently even a conservative court can be caught up in the counter-culture imputation that if something is "natural" or ubiquitous it is, of necessity, safe and, of course, good. (The same argument has been raised about estrogens and their addition to cattle feed. After all, how could estrogens be bad? They are present in every woman — and man — just more in women. Does the ambient level of hormones have anything to do with higher levels of breast cancer in women?)

However, laying aside the various arguments given for setting aside the Consumer Product Safety Commission order (including the bodies-in-the-streets argument "not a single case of urea-foam formaldehyde insulation cancer in humans was cited"), it may still be useful to look at the formaldehyde data. My colleagues and I have published an article[17] on formaldehyde and I draw liberally upon that work. There are three findings of consequence:

1. In 1982, there were approximately 1.3 million workers exposed to formaldehyde on their jobs. The arithmetic average exposure ("most probable exposure level") for the several industries ranged from 0.04 ppm (mean of 10 observations of workers as "producers of rubber and miscellaneous plastic products") to 8.3 ppm ("mean of 8 observations — area samples" in a group of biology instructors).

2. In seven epidemiologic studies (of varying quality and power) "of industrial populations in the United States and Great Britain . . . exposed mainly to formaldehyde or to multiple chemicals including formaldehyde . . . increases in cancer have been reported in six. . . . Cancers of the mouth, tongue and larynx have been reported in excess among persons exposed to formaldehyde." (As I read Siegel's tables, six of seven studies showed statistically significant increases in at least one site.)

3. Laboratory studies show a steep, nonlinear dose-response curve (for nasopharyngeal cancers in rats)

$$\text{Lifetime incidence} = 2.3 \times 10^{-4} \, (\text{dose in ppm})^3$$

cancers from which risk estimates could be made. I have read two criticisms (in the OSHA record) of attempts at risk assessment. One led me to wonder what the critic had read of the relevant papers, and the other was of the kind that reminded me of a remark by a University of California professor, "Academics can make a living picking at the margins. . . ." (Fairfax, S., 1982. "Old Recipes for New Federalism." Environmental Law, Vol. 12 No. 4, pp. 945–980.) To me, some critics seem to ignore ". . . what should be a principal objective of the . . . federal system — the promotion and protection of the autonomy and welfare of the individual." (". . . the federal structure guards one part of the society against the injustices of the other part" [James Madison, Federalist Paper No. 51]).

Table 4. Right-skewed Distribution of Exactly Measured Exposures

Exact Exposure (ppm)	Number of Employees
1	10,000
2	20,000
5	40,000
8	20,000
10	7,500
20	2,500

Average Exposure:	
Arithmetic mean	= 4.35 ppm
Geometric mean	= 4.246 ppm
Median: Interpolated	= 3.5 ppm
Exact	= 5 ppm
Mode (same as "exact" median)	= 5 ppm

Direct application of this formula implies a virtually safe dose (VSD) leading to a 1×10^{-6} lifetime risk, of 0.163 ppm, 1 with the 95% (lower) confidence limit dose as 0.0009 ppm. A direct NOEL approach based on doses that irritate the nasal mucosa found the VSD to lie between 0.01 and 0.001 ppm.

If we were to attempt to use the animal data to estimate risks for humans, there being inadequate dose-response data in humans, we must pay substantial attention to the fact that the dose-response curve (in rats) is not linear. The use of any average exposure (arithmetic, geometric, harmonic, median, mode, etc.) is very likely to underestimate the risk substantially. A former EPA science administrator, while noting that he had few data on general population exposure and hence could not compute risks, suggested the geometric mean as the "average" exposure to use—if he ever got enough exposure data upon which to base computations.[18]

A simple example can show why underestimating occurs. Say that an industry has a right-skewed distribution of exactly measured exposures—as illustrated Table 4. (Most exposures are likely to be right-skewed. The actual formaldehyde data Siegel has assembled is strongly right-skewed.)

Putting several of these values into the (rat) equation,

$$\text{Lifetime incidence} = 2.3 \times 10^{-4} \text{ (d ppm)}^3$$

gives figures as shown in Table 5.

Thus using any of these measures of central tendency for the dose in computation in the dose-response equation has the effect of underestimating the risk by a factor ranging from 3.4 to 10, i.e., it gives one-third to one-tenth the appropriate number of cases.

This example gives the estimate for only one hypothetical industry. If one uses actual industry averages for exposure, and then enters the estimating

Table 5. Right-skewed Distribution of Exactly Measured Exposures

Dose Measure (in ppm)		Estimated Lifetime Number of Cases Among 100,000 Employees	Ratio:Direct Computation Estimate — Estimate Based on Means, Mode or Median
Arith mean	4.35	1850	5.3
Geom mean	4.246	1760	5.6
Median (interpolated)	3.5	986	10.0
Mode (and exact median)	5.0	2875	3.4
Direct computation[a]		9869	—

[a] $2.3 \times 10 x^4$ $(10{,}000 \times 1^3 + 20{,}000 \times 2^3 + \ldots + 2{,}500 \times 20^3)$.

equation to do a "direct computation" for the country as a whole, industry by industry (each industry's population multiplied by the industry average exposure, raised to the third power, multiplied by 2.3×10^{-4}, and then summed across all industries), an underestimate is sure to be made. If this says average exposure is (often or sometimes) not appropriate (which it does say, if the dose-response curve is nonlinear), then what is the moral of the fable derived from the above table?

Moral No. 1: Exposure data in humans are necessary if risk estimates are to be made. If the dose-response relationship is nonlinear (especially if the response increases rapidly with dose), individual exposure data are essential. If we use averages, we will underestimate risk if exposure is right-skewed (which it almost certainly will be).

Q. What is the prospect for getting decent individual exposure data for materials that will not be suspected as being dangerous until tomorrow or the day after?

A. Nearly zero.

Q. Does this mean that no real risk assessments can be made?

A. No, but we will need to be realistic and not re-ific (coined word from same source as re-ify) about our computations. Just because an answer is numerical does not make it worth implicit belief (nor instant distrust, either.) Sometimes a "sense" of what should, or could be done, may be more important, more useful, more informative, and more accurate than a computation, particularly one based on luftwelt data. On the other hand, unceasing cynicism about data is no great help, either. Be critical? Yes. Be cynical? No.

Moral No. 2: Risk assessment needs careful assessment. Without assessment it is only risk. Scientists must not be King Canutes. We are subject to Supreme Court decisions as much as are administrators and lawyers. Yes, we are. Some of us are going to have to make risk assessments. We'll have to make our assumptions explicit. We'll have to talk about uncertainties. And we probably should say who paid us when we did our arithmetic. (I am now a Senior Science Advisor to Clement Associates, a science and engineering firm; a senior fellow at the Environmental Law Institute, a nonprofit research group; and a professor of biostatistics at the Uniformed Services University of the Health Sciences, the U.S. military medical school. By far the largest portion of my income comes from my federal government pension, subsequent to my work at the National Cancer Institute.)

SUMMARY

Can epidemiology help us make risk assessments? Yes, but not nearly as much as some people think it might. Perhaps risk assessment should be confined to helping set priorities for regulation, rather than for making absolute numerical estimates.[19] If we come to that conclusion, we are suggesting to Justice Powell that he retreat from his "Benzene" opinion as not in keeping with current science.

The major missing element is exposure information. I would hope that the marketing and sampling statisticians who have been so good at sampling other things would develop some ways to do good (sample) measures of exposures. And we need these on an individual basis, if dose-response is not linear.

All these problems (in doing a risk evaluation) were nicely summarized some time ago by the mathematician-philosopher Bertrand Russell: "So many circumstances of a small and accidental nature are relevant that no broad and simple uniformities are possible." But that does not mean that nothing can be done, or that what can be done is, by definition, bad science.

REFERENCES

1. Federal Register, 47(2), January 5, 1982.
2. U.S. Supreme Court. Review of Public Comments on the Reevaluation Through Rulemaking of the OSHA Cancer Policy. Clement Associates, Arlington, VA, August 31, 1982.
3. Interagency Regulatory Liaison Group. 1979.
4. Karstadt, M. et al. A survey of availability of epidemiologic data on humans exposed to animal carcinogens. In: Peto, R. and Schneiderman, M., Eds.Quantification of Occupational Cancer. Banbury Report No. 9. Cold Spring Harbor Laboratory, Cold Spring Harbor, New York, 1981.
5. National Academy of Sciences/National Research Council (NAS/NRC). Risk Assessment in the Federal Government: Managing the Process. National Academy Press, Washington, DC, 1983.
6. Nicholson, W. J. Dose-Response Relationships for Asbestos and Inorganic Fibers. Environmental Sciences Laboratory, Mt. Sinai School of Medicine, New York, 1981.
7. Schneiderman, M. A., Nisbet, I. C. T., and Brett, S. M. Assessment

of Risks Posed by Exposure to Low Levels of Asbestos in the General Environment. Bundesgesundheitsamt Berichte 4 pp. 3/1–3/27, 1981.

8. Seidman, H. et al. Short-Term Asbestos Work Exposure and Long-Term Observation. *N.Y. Acad. Sci.* 330: 61–90, 1979.

9. Newhouse, M. and Berry, G. Predictions of Mortality from Mesothelial Tumors in an Asbestos Factory. *Brit. J. Ind. Med.* 33: 147–151, 1976. (See also: Patterns of disease among long-term asbestos workers in the United Kingdom. *Ann. N.Y. Acad. Sci.* 330: 53–60, 1979).

10. Henderson, V. and Enterline, P. Asbestos Exposure: Factors Associated with Excess Cancer and Respiratory Disease Mortality. *Ann. N.Y. Acad. Sci.* 330: 117–126, 1979.

11. Dement, J. M. et al. Estimates of dose-response for respiratory cancer among chrysotile asbestos textile workers. Fifth International Conference on Inhaled Particles, Cardiff, Wales, 1980.

12. Breslow, N. et al. Multiplicative Models and Cohort Analysis. *J. Am. Stat. Assoc.* 78: 1–12, 1983.

13. Organization for Economic Cooperation and Development (OECD). Control of Toxic Substances in the Atmosphere: Asbestos. OECD, Paris, France (Document Env/Air 81–18), 1982.

14. Walker, A. Projections of Asbestos-Related Disease 1980–2009. Epidemiology Resources, Inc. Chestnut Hill, MA, 1982.

15. Nicholson, W. J. Comments on "Projections of Asbestos-Related Disease, 1980–2000" by A. M. Walker. Environmental Sciences Laboratory, Mt. Sinai School of Medicine, New York, 1983.

16. International Agency for Research on Cancer (IARC). IARC Monographs on the Evaluation of Carcinogenic Risk of Chemicals to Humans. Vol. 29. Benzene. World Health Organization, Lyon, France, 1982. (See also: *Science* 217: 914–915, Sept. 3, 1982).

17. Siegel, D., Frankos, V. H., and Schneiderman, M. A. Formaldehyde Risk Assessment for Occupationally Exposed Workers. *Reg. Toxicol. Pharmacol.*, 3: 355, 1983.

18. Todhunter, J. Internal EPA memo dated February 1982. Subject: Review of data available to the administrator concerning formaldehyde and di(2-ethylhexyl)phthalate DEPH: To Anne M. Gorsuch, through John W. Hernandez (see pp. 12–13 and 16), 1982.

19. Doll, R. and Peto, R. Quantitative Estimates of Avoidable Risks of Cancer in America Today. *J. Nat. Cancer Inst.* 66: 1191–1308, 1981.

SECTION IV

Defining Health Risks

CHAPTER 8

Engineering Perspectives from an Industrial Viewpoint

Patrick R. Atkins

INTRODUCTION

How many times has each of us said "If we lived in an ideal world . . ."? Unfortunately, the world is not ideal, and perhaps the place where that lack of logic and realism is most evident is in the environmental arena. There seems to be no part of the nation's environmental effort that is not fraught with controversy, emotion, dissatisfaction, and ill will. Miraculously, progress is being made, but it is painful progress and progress that leaves scars.

Recently, Dr. Lester Lave wrote in *The Washington Post* on the subject of the environment and how to develop solutions that will work in our complex society. He stated, "The main difficulty is that while danger lurks everywhere, it is nowhere very great. Pollution in the air, the water and the environment generally increases disease risks, including cancer. Yet rarely, if ever, are the risks high enough to constitute an emergency. For example, even at Love Canal, exposure to this stew of hazardous substances would be estimated to increase lifetime cancer incidence by perhaps one percent. I do not mean to disparage these concerns, but rather provide a rough estimate of the problems faced. Lifestyles, including smoking habits and diet, have a

J. B. Andelman and D. W. Underhill (Editors), *Health Effects from Hazardous Waste Sites*
© 1987 Lewis Publishers, Inc., Chelsea, Michigan — Printed in U.S.A.

much greater effect on disease rates, including cancer, than current environmental exposures."[1]

We live today in a society of individuals, each with abilities and capabilities to make decisions and judgments on an infinite variety of subjects. We encourage the exercise of those judgments and provide the mechanism for pluralism to flourish. With the advantages of this ability to act as individuals must come some disadvantages, and I would like to deal with some of those.

THE IDEAL WORLD

In that ideal world mentioned earlier, if an environmental situation were perceived to be a problem or a potential problem by society, a regulatory process would be initiated. The problems would be studied in detail, the facts developed, alternative solutions outlined and tested, and all parties affected would reach a consensus position on how to rectify the situation. Producers would recognize that the air, water, and land are not free resources to be used indiscriminately for waste disposal—thus externalizing costs in an effort to gain a competitive position. Society in general would recognize that there are risks associated with every facet of our lives, including the production of goods, and that these risks must be managed in a realistic fashion. The consensus solution would then be implemented in a timely fashion and in a cost-effective manner. In other words, the optimum solution would result from the collegial effort. So much for the fairy tale . . . back to the real world.

THE REAL WORLD

Our system works differently. We have evolved a litigious society that seems to be based on mistrust and misunderstanding. Almost any issue that deals with the environment or human health quickly becomes polarized, and the gulf widens with time rather than becoming more narrow. In order to discuss an issue, we find ourselves drifting rapidly to one of the extremes.

In the early 1970s, the cry was "Lake Erie is dead." Although there was some exaggeration, the issue was real. Attention was directed to a problem, action was initiated, and regulations were developed. Massive cleanup action was begun, the purpose was accomplished, and Lake Erie, along with innumerable other lakes and streams, was improved tremendously. Lake Erie was not dead, but the rallying cry worked.

Likewise, in the mid-1970s, the cry that the ozone layer was being destroyed by chlorofluorocarbons became popular—so popular in fact that an industry was essentially eliminated in the United States. The issue was very polarized, but the system worked—perhaps with a little too much pain—but it worked nevertheless. Almost as a postscript, the U. S. Environmental Protection Agency (EPA) announced in 1982 that further efforts to reduce the use of chlorofluorocarbons would be dropped because subsequent measurements and improved modeling techniques have shown the problem to be much less severe than was originally projected.

My point is that our system of regulatory development fosters mistrust and forces us to become polarized on almost all issues. Most regulation takes place in a crisis atmosphere, and decisions always seem to be needed immediately, invariably without adequate scientific information or a thorough understanding of the problem. Unfortunately, I don't see that situation changing. The complexity of the issues facing us in the environmental and public health arenas will only increase, exacerbating the regulatory problems. Our grandchildren will no doubt wrestle with the same problem.

As Betsy Ancker-Johnson said in a recent speech to the Stone and Webster Executive Conference, "In the real world, science and logic are only tangentially related to the regulatory process."[2] Most, if not all, the participants in the debates over complex issues are not equipped to gather, weigh, and understand the data, much less to objectively evaluate conflicting data and reach conclusions significantly different than their original premise. We all tend to look for or to the data that support our positions.

Nevertheless, we are blessed with the system that we have, and we must all learn to make it work as effectively as possible. Therefore, I believe it is critical to the success of our national hazardous waste management program for us to be able to accurately assess the health potential of waste disposal sites. It is a key part of the multiple step process that leads to good and effective regulations.

THE ROLE OF SCIENCE

There can be no substitute for a firm scientific basis for a regulatory action, and the health assessment is an integral part of that database. Without such information, effective compliance will be difficult to achieve: there simply cannot be enough regulators to ensure continuous compliance at every location, unless the regulated community also believes that the limitations are necessary.

A prime example of this problem was described in the April 27, 1983 issue of *The Wall Street Journal*.[3] The article describes the Salsbury Labs hazardous waste site near Charles City, Iowa, which was ranked by the EPA as the ninth most dangerous site in the United States. Yet, "Indeed, the striking fact about this rural community of 9,000 is how little has changed: None of the small businesses at the edge of the site—including the bowling alley, an automotive parts shop and an appliance store—have been vacated. No demonstrations have been held, no lawsuits filed and no citizens' committees have been formed. . . . More basically, people in this conservative heartland regard the environmental movement – and federal enforcers – as suspect." The new owners of the labs analyzed the hazardous waste problem at length before the recent acquisition and reached the conclusion that the problem wasn't so large that it "couldn't be handled by reasonable people." Yet, the site is listed as No. 9.

The EPA estimates that there are 15,000 uncontrolled waste sites in the United States today that must be cleaned up.[4] Some sites may cost 60 million dollars or more, and others will cost less than one million. If the sites average between three and four million dollars each, cleanup would cost the nation 50 billion dollars. Newsweek[5] recently estimated that this cost would be near 260 billion dollars. An OTA study states that annual expenditures for hazardous waste management, both onsite and offsite, were probably four to five billion annually at the beginning of the decade and will rise to 12.5 billion by 1990 (in 1981 dollars).[6] These costs represent 1 to 2% of the total sales of the chemical and allied products industry (assuming they are responsible for about half of all the hazardous waste generated). This is a significant drain, both in terms of cleanup and ongoing costs to the nation, and the companies paying these costs need good scientific information upon which to base decisions of such magnitude. As the Salsbury Lab site proves, the data are not conclusive.

However, the database is weak. I'll relate one experience that my company has had that illustrates this fact. A brief review of the literature and news reports leads me to conclude that there are numerous examples that could be used to demonstrate this point.

In late 1981, Alcoa was notified that a waste oil processor that had handled our wastes had gone bankrupt and left two hazardous waste sites that were causing significant harm to the environment. It was determined that one site was used exclusively for Alcoa wastes, and we agreed through the Consent Decree process to perform the necessary cleanup. During the period while we were developing the agreement, numerous stories appeared in the local press describing the hazardous nature of the site. Now that our contractor has been allowed on the site to conduct a true assessment and initiate a remedial action plan, we have learned that the site did not contain hazardous materials at all. There was no oil present, and the water could be discharged after rather simple treatment. The soils and sludges are still being analyzed, but I am confident that those materials will also prove to be nonhazardous.

This incident is, in my opinion, the classic case of overreaction that I described earlier and that The Wall Street Journal reported – overreaction that leads to polarization that leads to mistrust and misunderstanding. Scientific fact must be brought into the issues, and scientific facts must be the basis for decisionmaking.

ASSESSMENT METHODOLOGY

The EPA has developed a list of 419 priority sites throu
for which Superfund money could be provided. Yet, the
those sites were chosen is very nebulous. It appears t...
chosen, based on *potential* for environmental damage in the uncontrolled
state, rather than taking into account the impact of any remedial action that
had taken place at any of the sites. On this basis, almost all the selected sites
had groundwater contamination potential and 37% of the selected sites had
potential air quality impacts. The very simplified approach taken, though
obviously used because of the time pressures that existed, could result in
misleading statements and inappropriate priorities. On September 2, 1983,
the EPA proposed to add 133 additional sites to the original list and drop six
sites from that list because the degree of potential hazard was less than
earlier believed.

The assessment of hazard potential is complex and difficult. However, an
interesting approach and a better one for assessing the health impact of a
given hazardous waste site is to consider fugacity. Fugacity is the measure of
the escaping tendency of a compound in a given environment — or more
simply stated — the mobility of a substance in its environment. A good
example of why fugacity might be of significance is polychlorinated
biphenyl (PCB) contamination. PCBs tend to be rather soluble in certain
oils, relatively insoluble in water, retained by many types of soils, and not
very volatile at ambient temperatures. Therefore, PCB contamination from
a landfill, though complex, can be studied and quantified, given the right
information on fugacity. Likewise, PCB contamination of people follows
the same pattern: small amounts of contamination in air, more potential in
water, and the greatest potential in food. Therefore, the presence of PCBs
at a site will be of concern, but will not necessarily mean that the health risk
potential is great.

Estimating the environmental behavior of various chemicals can be done
by a three-step process that incorporates consideration of the fugacity of the
material and the environment in which the material is found. In the first
and least sophisticated level of examination, the potential for release of the
material should be compared with "allowable concentrations that can be
tolerated in the environment." These levels can be determined from existing
or proposed regulatory limits or estimated from the general health literature. If there is a possibility that any of these limitations might be
approached, then a second level of examination should be undertaken.

In the second level of examination, variables such as Henry's law coefficient, the solubility of the pollutant in available liquid streams and the
sorption coefficient of the material in contact with other materials in the
disposal site or the soils around the disposal site and the amount of material
present should be examined closely. Much of this information should be

available in the literature and records of the disposal site. Other information can be estimated through the use of various modeling techniques, i.e., accounting for changes in variables as a result of seasonal temperature changes.

In the third level of estimation, site-specific environmental information such as soil characteristics, groundwater information, wind conditions, landfill cover, etc., should be incorporated into the estimation of potential for release, and included in the detailed calculations of material movement in air, water, or soil.

It is important in understanding the impact of a hazardous waste disposal site on the environment to understand how the material is partitioned between the air and the water and the rocks and soil in the environment. A critical step in determining the environmental behavior of the site is to identify the source in terms of its magnitude and form, then to estimate the reactions that are occurring and the rate of transport of materials. Finally, that information can be used to determine if a hazard to the environment exists and what steps might be necessary to mitigate those impacts.

GROUNDWATER MONITORING

An effective health assessment program for a hazardous waste disposal site should include appropriate groundwater, surface water, and air quality determinations. The fugacity concept should be carefully considered when designing these systems. Since the data are complex and expensive to obtain, a well-designed monitoring system focused on the items of interest will serve much better than a broad-based, poorly designed program.

One of the areas where rapid progress is being made is in the measurement and modeling of groundwater impacts, although it is far from fully developed. A combination of monitoring well data and modeling information can lead to improved understanding of potential groundwater contamination and provide the basis for cost-effective steps to mitigate adverse impacts, and such studies should be conducted before conclusions are drawn about the potential impacts of disposal sites.

Current groundwater monitoring requirements call for at least one upgradient well and an appropriate number of downgradient wells. This concept is reasonable, but the complexities of the groundwater system in many locations make it extremely difficult to accurately assess groundwater quality in the short term. The EPA has prescribed statistical tests to be run on quarterly samples split four times in an attempt to increase the meaningfulness of the data. However, the test prescribed can produce very misleading results since the samples taken are not truly independent data points. A more realistic groundwater monitoring and data assessment program is needed before we are able to truly determine when and if a disposal site is adversely impacting the groundwater in an area.

A recent study has been conducted around an abandoned creosote facility in Conroe, Texas by the National Center for Groundwater Research at Rice University.[7] Field methods for water sampling were used in a characterization of groundwater contamination in a shallow aquifer underneath the site. Different drilling techniques were used for 21 bore holes and the ultimate installation of eight monitoring wells. After several years of monitoring, the plume extent and rate of migration have been established. The data indicate that chloride migration (inorganic pollutant) from the site is relatively extensive, reaching a well 300 feet from the site, while the organics have migrated at a much slower rate indicating high adsorption impact by surrounding soils. Such data are essential in establishing potential impacts and management strategies.

Kelly et al.[8] recently reported on efforts to assess and control groundwater contamination around a chemical dump in Rhode Island. At this site, a combination of groundwater data and modeling information was used to develop a better understanding of the degree of contamination and the rate of movement. Contamination levels near the site were calculated using analytical dispersion models, and those results compared favorably with trends and levels of observed contamination. The retardation coefficient for organic contaminants in the groundwater was estimated to be 3.25, and when used in the models, produced results that closely agree with observed data. A recharge/recovery system was then designed and field tested.

Similarly, Woodruff et al.[9] reported on an analysis of the impacts of a hazardous waste disposal site on the public water supply well field. The study examined the feasibility of ten proposed remedial actions for protection of groundwater supplies from contaminant migration from a hazardous waste disposal site. The public water supply well field was located two miles downgradient from a landfill where direct dumping of liquid onto the ground and burial of wastes in the landfill had occurred. The well field was used on a seasonal basis to supply a portion of the water for the city. To date, none of the wells has been contaminated; however, monitoring wells installed between the landfill and supply wells show that under present operating conditions, capture contamination at production wells in the upper aquifer may occur in the near future.

A good calibration of numerical models to existing data enabled simulations of proposed solutions to be made for thirty years in the future. The results of the modeling showed the plume migration characteristics for each alternative. Five alternatives were deemed worthy of further examination in terms of satisfying cost and noncost criteria. Future monitoring of the contaminant migration in both aquifers is needed, however, to verify the model predictions. These data and future studies will provide the opportunity for reasonable decisions to be made in a timely fashion to mitigate adverse impacts should they occur.

SURFACE WATER MONITORING

Surface water quality management at a site can be relatively simple or extremely complex. Groundwater mounds, improper caps, unusual topography, etc., can lead to surface water contamination from runon water, runoff water, or leachates. Here too, improved monitoring and testing procedures are needed. For example, the current leachate test prescribed by the EPA requires that the leaching fluid be maintained at an acid pH, presumably to simulate the type of liquid that is expected in a municipal landfill. Since many hazardous waste sites do not include municipal wastes, the standard leachate test does not give a valid picture of the potential for a site to impact the environment. To gain a true understanding of how a material will interact with the environment of a disposal site, more realistic methods of leachate characterization are needed.

AIR QUALITY MONITORING

Air emissions from hazardous waste sites are also poorly understood. It is surmised that hazardous materials can move from a site as:

- released dusts and vapors during transfer in the site
- contaminated dust generated by vehicle traffic
- volatilization from the landfill or lagoon
- droplets generated by wind action or mechanical spraying

Since disposal sites are low-level area sources, emission rate determinations and dispersion modeling to estimate downwind concentrations are difficult. In most cases, the fugacity concept would lead to the conclusion that the problem is small, but in some instances, certain compounds are of enough concern that an air monitoring program may be necessary.

Lipsky[10] reported on the development of a system to use air monitoring instrumentation at a waste disposal site to determine the characteristics of materials buried in the site. Portable air monitoring instrumentation is capable of providing data necessary to locate volatile components buried at disposal sites and to provide rapid screening of multimedia environmental samples to determine the volatile compounds present. Portable field equipment provides the opportunity for a large number of real-time samples to be taken so that the site can be characterized effectively. Monitoring of the soil gases provides information on the types of materials present and the boundary limits of contaminated areas that might contain volatile wastes. The method is based upon the fact that most volatile compounds move upward through soil toward the air–soil interface by molecular diffusion through the interconnected cracks in soil or spaces. The rate of diffusion is governed by many factors and is a complex function of soil type, porosity, temperature, physical and chemical characteristics, moisture, and other factors. The equipment used by Lipsky included:

1. *Organic Vapor Analyzer (OVA)*. This instrument equipped with a flame ionization detector provides a continuous reading of total volatile hydrocarbon concentrations in ambient air at the range of 1 to 1000 ppm. The instrument can be used to measure total hydrocarbons present or with the use of particular columns can serve as a gas chromatograph, allowing for the separation and identification of most common volatile organic compounds. The instrument is easily carried by one operator and can be operated on battery power.

2. *HNU Photoionization Detector*. The HNU device is a portable continuous vapor analyzer containing a photoionization detector which has the capability of detecting many organic and some inorganic compounds. The instrument must be calibrated against a benzene standard and is, therefore, only semiquantitative for nonmethane hydrocarbons. It serves as a good screening tool but cannot be used to quantitatively define specific hydrocarbon emissions.

3. *Photovac*. A Photovac device is an instrument that provides analysis of small samples injected by syringe into the chromatographic columns. Grab samples must be collected and injected by an operator. This system does not provide a continuous reading, but rather has the capability to detect volatile organic materials in a parts per trillion range in the grab samples analyzed. The instrument has the greatest usefulness as an ambient air sampling device to determine the concentrations of any volatile organic materials escaping a disposal site and migrating downwind.

Lipsky reports that the use of field instruments of this type offers the following advantages:

- equal or better sensitivity than some laboratory methodologies
- rapid turnaround time
- processing of large numbers of samples to provide a maximum amount of data
- flexibility to change sampling locations and analytical needs as warranted by the data

This represents a major advance in the capability to determine potential for airborne environmental impacts from hazardous waste disposal facilities.

Similarly, Schmidt and Balfour[11] described a method to directly measure gas emissions from hazardous waste sites using isolation flux chambers and electrochemical or chromatographic detectors. Flux chambers are designed to cover a specified area (several hundreds of square centimeters) with an inert chamber that does not impact the natural flux of gaseous emissions from the disposal site surface. Special care is given to the chemical inertness, the structural integrity, and light transmission of the chamber, and the system is designed to provide the appropriate amount of sweeping air to remove the pollutants generated without affecting the hydrocarbon flux rate. The data collected suggest that fluctuations in gaseous emission rates occur and may be related to several factors including: natural emission processes, temperature, sky cover (ultraviolet radiation), surface distur- bances, etc. Data from surface measurements from properly selected loca- tions at a waste site in conjunction with a knowledge of the variability of the emissions and the uncertainties associated with the emissions measurements can provide a tool for characterization of the total emissions at a waste facility for both undisturbed and disturbed surfaces.

In addition, ground probes can be used to obtain cost-effective, shallow subsurface emission data over a large area. The advantage of this type of a monitoring program is that the emission measurement system can be uti- lized at a specific location under a known set of environmental conditions to determine the emission rate from a disposal site. Samples can be repeated under a variety of conditions to gain a better understanding of the impacts of various parameters on emission rates from a given source. By integrating the information available, a disposal site can be effectively characterized to account for the variety of influences that impact emission rates.

Data on emission potential from disposal sites can be used to provide the source terms for dispersion models to predict downwind impacts from dis- posal sites. Such models are extremely difficult to develop and even more difficult to verify. The concentrations under consideration are extremely small, and the impacts from low-level releases such as landfill sites are significantly influenced by minute topographic parameters and complex wind patterns near the ground's surface. A number of attempts are under- way to develop better models so that reliable predictions can be made.

Hwang[12] reported on field evaluations of volatilization rate models for land disposal facilities. Currently available models utilize extensions of the elevated point source models that have been developed over the years for releases of airborne contaminants from process sources. The assumptions associated with the application of such models to the low-level, area releases from waste management sites have significant impacts on the predictions and, therefore, must be considered carefully. Some studies have attempted to relate predicted ambient concentrations to measured levels with marginal success.[13] These studies will continue, and as the tools become better defined, the potential impacts of waste disposal sites on ambient air quality will be better understood.

Air quality programs may be of greater importance when consideration is given to cleanup activities at a site. Excavation, loading, and transport of materials in a landfill can be a significant, though short-term source of emissions; and the potential impacts of such activities should be carefully assessed before a cleanup begins. It may be more realistic to properly contain the material onsite rather than attempt to move the wastes. Again, fugacity is a critical consideration.

However, because of the complexity of the problem, the impacts of hazardous waste site cleanup activities on air quality are often not completely assessed as decisions are made concerning site cleanup activities and mitigation and management programs. This consideration should be a part of the overall risk assessment conducted when a long-term hazardous waste management program is developed for a particular disposal site. As improved predictive tools are developed, the ability to utilize such impact information in cleanup decisions will make it possible to also consider this important aspect of the pollution problem.

CONCLUSION

My conclusion is twofold—the regulatory process must be correct to be effective and the scientific basis for risk assessment is of paramount importance. Regulations are often arbitrary, uninformed, and inefficient. Such regulations cannot be effectively enforced. They are costly, and most damaging of all, they do not accomplish their goals. This is certainly the case in the realm of hazardous wastes. We must learn to better assess the risks associated with hazardous waste sites. The programs must be better designed and conducted in time frames that will allow appropriate judgments to be made, based on fact rather than emotion. The problem is real, and we must all work together to develop the procedures and methodologies to gather good scientific data upon which to base decisions that can severely impact us all.

REFERENCES

1. Lave, Lester. Environmental Solutions Aren't Simple. *The Washington Post*, p. A19, April 6, 1983.
2. Ancker-Johnson, B. Taking Our Medicine. *The Environmental Forum*, April, 1983.
3. Jasen, G. Iowa Town Learns to Live with Chemical Waste. *The Wall Street Journal*, April 27, 1983.
4. Conservation Foundation Letter. Clean-up and Fix-up Costs Rise Relentlessly. March, 1983.
5. The Toxic Waste Crisis. *Newsweek*, March 7, 1983.
6. Office of Technology Assessment. Technologies and Management Strategies for Hazardous Waste Control. March, 1983.
7. Rodgers, A. C. and Bedient, P. B. Ground Water Transport and Attenuation from a Waste Site. Proceedings of the ASCE 1983 National Conference on Environmental Engineering, Boulder, CO, p. 548, July, 1983.
8. Kelly, W. E., Powers, M. A. and Virgadamo, P. P. Ground Water

Pollution: Assessment and Control. Proceedings of the ASCE 1983 National Conference on Environmental Engineering, Boulder, CO, p. 556, July, 1983.

9. Woodruff, D. A., Hughto, R. J. and Harley, B. M. Analysis of the Impacts of a Hazardous Waste Disposal Site on a Public Water Supply Well Field. Proceedings of the ASCE 1983 National Conference on Environmental Engineering, Boulder, CO, p. 751, July, 1983.

10. Lipsky, D. Buried Waste Characterization Using Air Monitoring Instrumentation. Proceedings of the ASCE 1983 National Conference on Environmental Engineering, Boulder, CO, p. 861, July, 1983.

11. Schmidt, C. E., Balfour, W. D. and Cox, R. D. Sampling Techniques for Emissions Measurement at Hazardous Waste Sites. Proceedings from the Management of Uncontrolled Hazardous Waste Sites Conference, November, 1982.

12. Hwang, S. T. Field Evaluations of Volatilization Rate Models for Land Disposal Facilities. Proceedings of the ASCE 1983 National Conference on Environmental Engineering, Boulder, CO, p. 680, July, 1983.

13. McCord, A. T. and Hanchak, M. J. Practical Applications of a Model of Emissions. Paper presented to the ASCE 1983 National Conference on Environmental Engineering, Boulder, CO, July, 1983.

CHAPTER 9

Lessons from Love Canal: The Role of the Public and the Use of Birth Weight, Growth, and Indigenous Wildlife to Evaluate Health Risk

Beverly Paigen and Lynn R. Goldman

ROLE OF THE PUBLIC IN DEFINING HEALTH PROBLEMS

In defining health problems at hazardous waste sites and the role of citizens and public interest groups, it is not a question of whether the public should be involved. They are involved and will remain so. In almost every hazardous waste situation where health studies are being done, it was the citizens living near the waste site who brought the problem to the attention of professionals. Perhaps some day in the future, government agencies will logically and systematically assess the hazardous potential of toxic waste sites and initiate studies of those having the greatest public health hazard. In the foreseeable future, however, health studies will start because local citizens initiate the chain of events that lead to the study.

The hazardous waste problem has led citizens to initiate action. It is in their own backyards that the waste is buried; it is their health and the health of their neighbors that are affected. Citizens are now organizing on local, state, and national levels to address these problems.

A national organization, The Citizens' Clearing House for Hazardous

J. B. Andelman and D. W. Underhill (Editors), *Health Effects from Hazardous Waste Sites*
© 1987 Lewis Publishers, Inc., Chelsea, Michigan – Printed in U.S.A.

Waste (Box 7097, Arlington, VA 22207), grew out of Love Canal. It publishes a newsletter, "Everyone's Backyard," with a rapidly growing circulation which is now 3000. This organization aids citizens living near hazardous waste sites, providing scientific information on the toxic effects of chemicals, information on landfilling technology, and alternatives to burying toxic wastes. It is also a resource center, linking people to relevant government agencies, providing methods of community organization, and referring to sources of scientific and legal help. In its first five months, the Citizens' Clearing House has had over 2000 phone calls and has visited 63 sites. In April, 1983, it held a leadership training conference in Ohio attended by citizen groups from 21 different hazardous waste sites.

The Citizens' Clearing House recognizes that health studies are best planned and carried out by public health professionals. Sometimes, however, government agencies do not have funds to carry out studies. Other times, they are skeptical and unconvinced that health studies are needed. Thus, the burden of proof is placed on the citizens, who attempt to carry out health studies themselves. One of the most frequent requests to the Citizens' Clearing House has been for a sample health questionnaire. In response, the Clearing House has prepared a community health profile to collect information about demographics and health problems. This systematic documentation can be used by communities more reliably than anecdotes.

The introduction of this method of data collection should be very useful. Most epidemiological studies do not discover new associations; they usually prove or disprove hypotheses. These hypotheses are usually suggested by workers, physicians, or the public. If a community collects information on demographics and a health profile, this should enable public health professionals to plan a more focused study.

Collection of health data by nonprofessionals is a phenomenon that is likely to become more common. A new book, *The Health Detective's Handbook*, explains in layman's terms many of the basic tools of epidemiology.[1] Edited by Dr. Marvin Legator, it contains a sample questionnaire, explanations of how to use birth and death certificates, and methods of data analysis.

When a public health professional begins a health study of the hazardous waste sites, he or she is faced with an unusual situation for a scientist. There is no possibility of doing a double blind study. The study participants know about their exposure. They also have a fairly clear idea of the hypothesis being tested and most likely have suggested the hypothesis themselves. The participants who live near the hazardous waste site have probably been concerned about their health, and have been educating themselves about the known toxicities of the chemicals which have been found on the site. Their recall of medical problems is likely to be excellent and therefore better than that of a control group. Furthermore, the community will want to be involved in many aspects of the study. Such a situation presents several challenges to the health professional. We would like to address the remaining portion of this chapter to two specific areas: (1) working with the community and (2) dealing with the difficult problem of bias in such situations.

WORKING WITH THE COMMUNITY

It is not a question of whether the community will be involved in the health study; it is a question of how. The relationship should be a positive one with the community input being a valuable asset to the study. However, if the health professional keeps the community at arm's length, in order to make the study more objective, the relationship could turn into one of distrust. At Love Canal and other sites, the relationship between professionals and the community seriously deteriorated, resulting in mistrust of government, damage to the image of scientists, and verbal attacks on well-intentioned public health professionals.[2-3] These problems can be reduced if two practices are followed. First, scientists should set up an advisory board of citizens and outside scientists. The board should advise in defining the questions to be answered, reviewing study design and protocols, interpreting data, and finally presenting the results to the community. This practice has worked in California. (See Chapter 10.) Second, some money should be set aside so that the community can obtain independent expert advice on study design and data analysis. There is a precedent for this: Love Canal, where the State of New York gave residents money to hire a toxicologist to monitor the safety plan and to interpret data to the community. This person served as a conduit of information and was able to resolve many controversies before they became serious.[2]

DEALING WITH THE PROBLEM OF BIAS

Since surveys of disease prevalence in a community near a hazardous waste site have problems with reporting bias, more objective measurements are needed. In 1980, Dr. Joseph Highland, then with the Environmental Defense Fund, and Dr. Beverly Paigen drew up a list of some possible objective measures.[4] We have now gained some experience with three of these: wildlife, birth weight, and growth at Love Canal. I would like to review the results briefly.

One possibility considered was the examination of health of indigenous wildlife. Rats are useful for hazardous waste sites in urban areas; meadow voles, *Microtus*, are useful for suburban or rural areas. *Microtus* live in grassy, moist areas and can be caught with ordinary mouse traps or trapped live. The measurements that can be made include population density, body and eye lens weight (to estimate age and maturity), organ weights as a function of total body weight, histology of key organs, and a variety of biochemical measurements. Reproductive history can be examined by the percentage of pregnant females and number and size of fetuses. In nonpregnant females, examination of the number and shape of placental scars permits evaluation of previous litter size and the number of fetal resorptions.

Another objective measure is the birth weight of human babies. These data are both objective, being available in vital statistics records, and sensitive to many environmental influences.[5-6] The frequency of low birth weight babies, those weighing less than 2500 gm, is about 7% in the normal population. Three hundred births in the exposed population would be required to detect a doubling in the frequency if α is 0.05 and β is 0.20.[7] If birth weight is used as a continuous quantitative variable, it should be more powerful statistically and be able to detect a difference with less than 300 births.

A third objective measure that can provide an early warning for exposure to toxic chemicals is the growth of children. Growth of children is a regular and predictable process for which extensive normative data exist.[8] Growth of children is affected by a variety of environmental factors,[8-9] including air pollution,[10] radiation,[11-15] and chemicals.[16] Measurements of children's growth are noninvasive, inexpensive, and potentially quite useful in evaluating hazardous waste sites.

Most hazardous waste sites contain neurotoxins, and exposed communities frequently mention neurological symptoms as health problems. Several objective tests for peripheral or central nervous system pathology are available.[17-18] These are more reliable than attempting to deal with reports of subjective symptoms such as dizziness, fatigue, blurred vision, and numbness of fingers and toes.

Other objective tests on blood and urine samples reveal several types of specific organ damage to the liver, kidney, bone marrow, or immune system.

INDIGENOUS WILDLIFE

Trapping native wildlife in a hazardous waste area has several advantages. It sidesteps the problem of human bias. Animal studies are relatively low cost compared to epidemiologic ones. Levels of toxic chemicals can be measured in body fat, and observed pathology can provide clues as to the types of toxicity occurring in the human population.

At Love Canal, the field vole, *Microtus*, was trapped for two years. Trapping was done in three areas: around the perimeter of a fence, approximately 200 m from Love Canal, which enclosed the canal and two rows of homes (Area I); in the Love Canal neighborhood along Frontier Avenue (Area II); and in a control area (0.4 to 2 km) (Area III). Population density was determined by trapping for three successive nights with traps set in an equally spaced grid. The average catches per 100 trap nights were 1.8 in the Love Canal area, 5.0 in Area II, and 8.7 in the control area (Table 1). The

Table 1. Population Density of Voles

Area	Trap Nights	Number of Animals	Population Density[a]
Area I—Love Canal	4093	73	1.8
Area II—Near Love Canal	1457	73	5.0
Area III—Control	1978	172	8.7

[a]Population density is expressed as the number of voles caught per 100 trap nights during the three days of trapping. The above data represent the average of three separate determinations of population density in fall of 1979, spring of 1980, and summer of 1980.

population density of 1.8 is considered to be very low for Microtus. In order to obtain a sufficient number of animals for statistical purposes, many more traps had to be set each night in Area I compared to the others, and more nights of trapping were done. All animals trapped were examined for maturity, sex, pregnancy status, body weight, and weight of key organs.

Mean body weight was lower for animals trapped in the Love Canal area than for control animals. This smaller mean weight was due to a lower percentage of large animals, suggesting that the Love Canal voles did not live as long. In order to determine the age distribution of this vole population more precisely, the dry weights of eye lenses were used to estimate age.[19] The age distributions were used to construct survival curves (Figure 1). Animals from the Love Canal area had the steepest survival curve, indicating that their lifespans were reduced. Based on these survival curves, the average lifespans after weaning were 24 days, 29 days, and 49 days in animals from Areas I, II, and III, respectively. Thus the lifespan after weaning in animals from the Love Canal area was half that of control animals. Reduction in lifespan was most pronounced in females. The expected percentage of females in each age group is approximately 50%, and this was seen in populations from Areas II and III (Figure 2). However, in Love Canal animals, the percentage of females drops sharply with increasing age, indicating that females have increased mortality. The reason is unknown, but we speculate that females may succumb to infection or to hemorrhage after giving birth.

Much more was learned from this study. Involution of the thymus in young animals from the Love Canal area suggested an effect on the immune system. Reduced weights of testicles and seminal vesicles suggested delayed sexual maturation in males trapped in the Love Canal area. Examining mature females for pregnancy status and for number and size of embryos or number of dead and resorbing embryos would have given information on reproduction, but there were not enough mature females in the Love Canal areas to show statistically significant differences. Lindane, dichlorobenzene and other chemicals were found in Love Canal animals, using fat analysis by GC/MS. Histological examination of organs is continuing, particularly

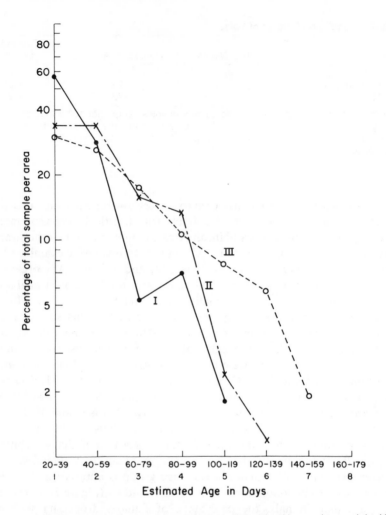

Figure 1. Distribution of voles by estimated age as determined by eye lens weight. Voles by both sexes trapped in different areas are combined by Area I, Love Canal ●———●; Area II, near Love Canal X———X; and Area III, control O———O. Numbers are adjusted for equal trap effort and expressed as percentage of total voles from each area. No voles less than 20 days old, the age of weaning, were trapped. Older voles who survived the winter and were trapped in early spring were eliminated from the analysis.

in the livers and adrenal glands, which were significantly smaller in Love Canal females. Further details of this study are presented elsewhere.[20]

At other sites, cross-sectional data on population density, body weights, organ weights, sex, maturity, and pregnancy status could be gathered in a month of trapping. More extensive studies could be done with indigenous

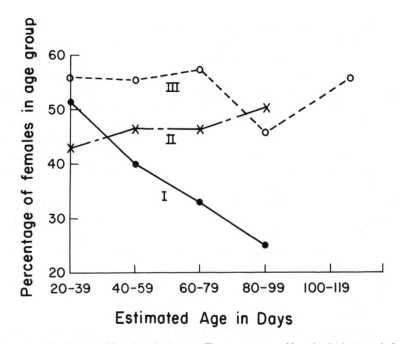

Figure 2. Distribution of female voles by age. The percentage of females in the sample for each age category are plotted by area: Area I, Love Canal ●———●; Area II, near Love Canal X———X; and Area III, control O———O. Overwintered voles are excluded.

wildlife by live trapping animals for reproductive and other studies in the laboratory. Laboratory-raised animals could be returned to contaminated areas in pens and sampled at intervals to do cohort studies.

Study of wildlife is a useful and important part of the evaluation of a hazardous waste site. However, routes of exposure are quite different for animals that burrow in the soil and eat local vegetation so that type and amount of exposure may differ significantly from exposure to humans. Moreover indigenous wildlife studies will not be suitable for those sites where the major route of exposure is via contaminated drinking water. Thus, studies on humans are also important.

STUDY OF CHILDREN LIVING IN THE LOVE CANAL NEIGHBORHOOD

Some results from a study on Love Canal and control children will now be discussed, focusing on birth weight and height of children and comparing these objective measures to some more subjective data obtained from interviews. These studies were done in collaboration with Joseph Highland, Mary Magnant, and Ted Steegmann.

Public groups played an important role in this study. When the citizens from Love Canal became disillusioned with the studies being done by the New York State Department of Health, primarily because of the veil of secrecy concerning the results of those studies, they approached the Environmental Defense Fund, a public interest group. The Environmental Defense Fund agreed to raise money and conduct health studies at Love Canal. The traditional sources of money for health studies were not available for this controversial situation, so again the public sector in the form of small private foundations played a major role. The first to give a grant for this study was the Gannett Foundation. The Gannett newspaper chain owns the *Niagara Gazette*, the newspaper that first broke the Love Canal story, and so this foundation had a local interest in the plight of the Love Canal residents. Once the Gannett Foundation awarded the grant to the Environmental Defense Fund, the study began. Additional funding came from the Ruth Mott Foundation, the Bingham Foundation, and the Environmental Protection Agency.

First a description of the Love Canal neighborhood and the study design is appropriate. The Love Canal neighborhood, which is on the eastern edge of the city of Niagara Falls, New York, has three natural boundaries: the Niagara River on the south, open fields on the east, and Bergholtz Creek on the north. Since no natural boundary exists on the western side, this edge of the Love Canal neighborhood was defined as 93rd Street. The neighborhood contains 1,115 dwelling units: 811 single family homes and a 304-unit government-subsidized apartment complex known as LaSalle Development.

Because the population in single family homes and apartments differed in race and socioeconomic class, two separate comparison groups were chosen. Other government subsidized housing units in the city of Niagara Falls were examined as possible controls for LaSalle Development, and Unity Park was chosen as the unit most similar to LaSalle with respect to racial composition and family size. Children residing in LaSalle are referred to as renters and those in Unity Park as renter controls.

For the second comparison group, the 1970 census information was examined for census tracts that were most similar to the Love Canal area containing single family homes in income, education, and percentage employed in manufacturing. The area chosen consisted of portions of census tracts 221 and 223 as well as part of the census tract containing Love Canal. Census tracts 221 and 223 are downwind and closer to the chemical manufacturing complex located in Niagara Falls than is the Love Canal area. Those residing in the part of the Love Canal neighborhood with single family homes are referred to as homeowners and those in the control census tracts as homeowner controls, even though some families did rent rather than own their homes.

Because of the close proximity of the chemical industry and the fact that western New York has over 200 known hazardous waste sites, the control areas were examined for hazardous waste sites using both visual inspection and a map of hazardous waste sites prepared by the New York State Interagency Task Force. None were found.

Exposure to chemicals was not equal for everyone in the neighborhood. Since no data on chemical levels were available to us, two crude indices of exposure were developed. One used distance from the canal as a measure of exposure by grouping homes into 120-m wide bands. This exposure index was similar for homeowners and renters. The analysis of distance is complicated by the fact that children who lived closest to the canal were evacuated two years prior to this study.

The second exposure index used the proximity of homes to probable preferential paths of chemical migration. The neighborhood in the past had several streams and swales, as well as a pond and some swamps. These have been filled with building rubble, providing paths for migration of chemicals in a predominantly clay area. The homes that are adjacent to historically wet areas are referred to as wet homes. Homeowners had far more exposure by this route than did renters. It had been suggested that health effects such as adverse reproductive outcomes occurred more frequently in homes built immediately adjacent to these former streambeds.[21-23] The determination of historically wet areas was made by soil scientists under contract to New York State. The Department of Health provided a map showing which homes in the section of the neighborhood east of Love Canal and south of Colvin Boulevard were counted "historically wet" in the Vianna et al. study.[22] This map also depicted the location of the swales and creeks in the remainder of the neighborhood. We assigned each home immediately adjacent to a swale or stream bed as historically wet.

Once the geographic area to be studied was selected, we attempted to obtain the participation of every child between the ages of 18 months through 16 years living in the area. A total of 963 children were in the study: 523 Love Canal and 440 control children distributed about equally by homeowners and renters, sex, and race. The response rate was about 62% for Love Canal homeowners, 63% for homeowner controls, 82% for Love Canal renters, and 81% for control renters.

Several kinds of biases could enter into this study. The first was selection bias. Did those children who did not participate differ from those who did? We randomly selected one-third of the nonparticipants and contacted them by letter, telephone, or personal visit to administer a short questionnaire. We found that the nonparticipants were similar in education and income, but differed in that children tended to be somewhat older. Nonparticipants were asked whether their children had the nine health problems most prevalent in the participants (Table 2). The percentage of positive responses for Love Canal nonparticipants was less than for participants, but the difference was not significant. In contrast, the control nonparticipants had significantly fewer positive responses for health problems than the participants,

Table 2. Comparison of Participants and Nonparticipants, Homeowners

	Exposed		Control	
	Participants	Nonparticipants	Participants	Nonparticipants
Number of children	239	27	259	44
Total n of positive responses	277	20	182	13 $p < .005$
Total n (n of children × 9 health questions)	2151	243	2331	396
Mean positive responses	12.9%	8.2%	7.8%	3.3%
Asthma	15	5	12	5
Abdominal pain	30	2	15	0
Seizures	13	0	3	1
Hyperactivity	23	4	10	2
Allergy to medication	28	0	18	0
Urinary tract infections	11	1	11	1
Skin rashes	81	4	67	1
Eye irritation	29	3	15	0
Learning problems	47	1	31	3

so that the study could have been biased in the direction that controls who had health problems were more likely to participate. Thus any true difference between Love Canal and control children would have been reduced.

Another type of bias would occur if exposed and control children were misclassified. A detailed residential history was obtained for each child including the gestational period. Children were classified as exposed or control depending on where they were living at the time of the survey. However, 12 Love Canal residents had lived there less than six months so exposure was minimal. Many children classified as controls had exposure to Love Canal chemicals; 25 had spent part of gestation in the Love Canal neighborhood, 20 had been born there, 48 had lived there, several regularly visited friends and relatives in the Love Canal area, and some had lived near other hazardous waste sites such as Hyde Park. If controls who participated were more likely than nonparticipants to have had some personal connection with hazardous waste, this would have biased the study, reducing any true difference between Love Canal and control children.

For the analysis of birth weight, we reassigned children to exposed and control groups by where their mothers lived during gestation. About 40% of Love Canal children and 5% of control children belonged to the group exposed during gestation. While 214 Love Canal and 25 control children were exposed during gestation (total 239), 707 children were left as controls (for 17 adopted, foster, or stepchildren, residence during gestation was unknown). Exposed and controls were well matched for race, sex and household income levels (Table 3). Both groups were about 60% white, 50% male, and had a median annual income of $15,000/year. The exposed group did differ significantly in that mothers of exposed children were

Table 3. Demography of Love Canal and Control Children by Residence During Gestation

	Exposed	Controls	p-Value[b]
Number[a]	239	707	
Race (% white)	61.1	60.8	NS[c]
Sex (% male)	49.8	49.2	NS
Median income category	$9–15,000/yr	$9–15,000/yr	NS
Percent mothers completed high school	64.8	73.5	<.01

[a]There were 17 children with residence during gestation unknown.
[b]Probabilities by Chi Square statistic with Yate's continuity correction.
[c]NS = not significant.

Table 4. Pregnancy Histories of Love Canal and Control Mothers by Residence During Gestation

	Exposed	Controls	p-Values
Smoking ≥ 1 PPD	20.5	20.8	NS[a]
Drinking ≥ 1 time/day	2.5%	1.4%	NS
History of miscarriage(s)	31.9	29.0	NS
History of stillbirth(s)	5.4	4.1	NS
Any medication during pregnancy	21.3	21.6	NS
Any illness during pregnancy	7.1	9.9	NS
Mean parity	3.4	2.7	<.01[b]
Mean maternal age at birth	26.3	24.9	<.01[b]

[a]NS = not significant.
[b]Probabilities by paired T-test.

somewhat less likely to have finished high school, had higher parity, and were slightly older at the time of delivery.

Detailed pregnancy histories were obtained. There were no statistically significant differences between exposed and control in smoking, alcohol consumption, history of miscarriages, history of stillbirths, use of medications during pregnancy, or illnesses during pregnancy. Medications were broken down into eight separate categories and illnesses into six categories, and no statistically significant differences between exposed and control were observed in any category (Table 4).

Exposed and controls were examined for the prevalence of babies who were low birth weight, that is who weighed less than 2500 gm at birth. Of the 210 children born in the Love Canal area for whom birth weight information was available, 12.3% weighed less than 2500 gm at birth compared to 8.6% of 697 children born elsewhere (p = 0.14, not significant). When low birth weight was examined separately for the homeowner and renter groups, the excess percentage of low birth weight babies was seen primarily in the homeowner group (Table 5). Multiple logistic regression analysis was used to control for race, sex, socioeconomic class, maternal age, parity, smoking, and other prenatal exposures. Results were used to provide an adjusted odds ratio. For renters there was no difference between exposed

Table 5. Low Birth Weight and Prematurity Among Love Canal and Control Children

	Exposed		Control		Adjusted Odds Ratio[c]
	N	%	N	%	
Birth weight <2500 gm[a]					
Homeowners	126	11.1%	356	4.8	3.0 (1.3–7.0)
Renters	94	13.8%	341	12.6%	1.1 (0.5–2.3)
Term <38 weeks[b]					
Homeowners	130	18.5%	351	15.1%	1.5 (0.9–2.6)
Renters	96	16.7%	353	15.3%	1.2 (0.6–2.3)

[a]N = 917. There were 29 children with birth weight missing.
[b]N = 930. There were 16 children with weeks gestation missing.
[c]Adjusted odds ratio derived from logistic multiple regression parameters (see text). 95% confidence interval is shown.

and controls. For homeowners, residence in Love Canal was strongly associated with low birth weight, giving an adjusted odds ratio of 3.0 (Table 5). As mentioned earlier, the homeowners had much more exposure by proximity to paths of preferential migration. The excess percentage of low birth weight babies was mostly due to residence in these historically "wet" homes. Distance from Love Canal had very little effect.

Low birth weight could result from premature birth or from being small for gestational age. In order to determine which of these alternatives was the most likely explanation for low birth weight, we corrected for gestational age. Respondents were asked how many days early or late the delivery of each child occurred. Days were rounded off to the nearest week and added or subtracted from 40. Children born at or before the 38th week of gestation were defined as premature. For both homeowners and renters, a small but not statistically significant increase in prevalence of premature births occurred (Table 5). Thus prematurity was not a major factor in the increase of low birth weight babies.

Decreased birth weight appeared to be primarily the result of retarded intrauterine growth as illustrated by a plot of mean birth weight of males by each week of gestational age (Figure 3). Love Canal boys are smaller than controls at each week of gestational age until forty weeks gestation. These birth weights were determined for children that participated in the study, so these data may be somewhat biased particularly at early ages by exclusion of premature babies that did not survive.

In order to determine whether the chemicals affected stature, children were measured under supervision of a physical anthropologist. The heights of the parents were also obtained to control for midparental height. Height was measured with a stadiometer. Slight upward pressure was exerted on the mastoid processes to straighten the spine and eliminate diurnal variation. Height was converted into percentiles for age using national statistics. Parental heights of Love Canal and controls were not significantly differ-

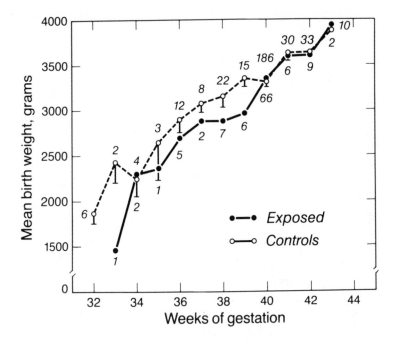

Figure 3. Birth weight of males at each week of gestational age for Love Canal and control children.

ent. Children who were both born in the Love Canal area and who grew up there, defined as having spent at least 75% of the child's life in the area, were significantly shorter (Table 6). Mean stature age percentile was 53.0 for controls and 46.6 for those both born and raised in the Love Canal area, a difference significant at p< .02. The other two groups were not different from controls, so evidently both gestational and childhood exposure are involved in this effect on stature.

These are the results we obtained from objective measures on the child's birth weight and height. Both showed an influence of being born near Love

Table 6. Mean Stature-age Percentiles (SAPR) by Residence at Birth and During Childhood

	N	Mean SAPR ± SE
Birth +, ≥ 75% Life Love Canal	170	46.6 ± 2.2[a]
Birth +, < 75% Life Love Canal	41	55.3 ± 4.8
Birth −, ≥ 75% Life Love Canal	85	53.5 ± 2.2
Birth −, < 75% Life Love Canal	610	53.0 ± 1.1

[a]Significant at p< .01 by student's T-test.

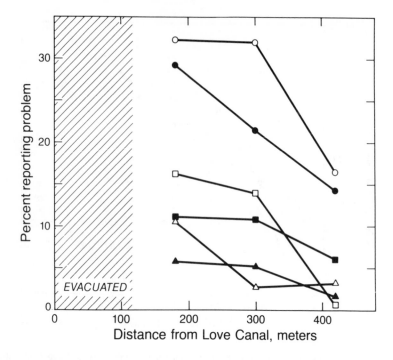

Figure 4. The effect of distance from Love Canal on the prevalence of health problems. Health effects shown are learning problems ○———○; skin rashes ●———●; eye irritation □———□; abdominal pain ■———■; hyperactivity △———△; and seizures ▲———▲. The area within 120 m of Love Canal (shaded) had been evacuated in 1978 and those children were not included in this study. The total N in each group was from 120–240 m, 166; from 240–360 m, 81; and from 360 to 480 m, 36. N was smaller for learning problems, which was asked only for school-aged children, and hyperactivity, which is shown for homeowners only.

Canal. These results can be compared with the results of the interview on health problems. Parents were asked whether a physician had diagnosed a number of illnesses. The analysis involved controlling for confounding variables by multiple logistic regression techniques to compute adjusted odds ratio. For most health problems, there was no difference between Love Canal and control children. There were no health problems for which controls had an elevated prevalence. Table 7 shows uncorrected prevalences and adjusted odds ratios for seizures, learning problems, hyperactivity, skin rashes, irritation of eyes, and abdominal pain. Adjusted odds ratios varied from 1.5 to 3.0.

There was a dose response of health problems with chemical exposure, using the crude indices of exposure. All of the health problems except abdominal pain were more prevalent in wet area homes compared to dry. In

Table 7. Percent Reporting Illnesses by Residence

	Exposed		Controls			
	N	%	N	%	Adjusted Odds Ratio[d]	
Seizures	520	5.2	440	1.8	2.4	(1.0–5.8)
Learning problems[a]	422	28.2	301	19.3	1.5	(1.0–2.2)
Hyperactivity[b]	238	9.7	259	3.9	3.0	(1.3–6.7)
Skin rashes	520	27.2	440	18.4	2.2	(1.4–3.5)
Eye irritation	520	12.7	440	6.4	2.0	(1.2–3.2)
Abdominal pain[c]	487	11.7	391	6.1	2.1	(1.2–3.5)

[a]Includes only children $\geq$ 60 months.
[b]Includes only homeowner children (see text).
[c]Includes only children > 36 months.
[d]Adjusted odds ratio derived from logistic regression parameters (see text). 95% confidence interval is shown.

addition all six health problems showed a gradient of prevalence by distance from the Canal (Figure 4).

To summarize, differences at Love Canal have been demonstrated by three objective measures: indigenous wildlife, birth weight, and stature. The results on birth weight confirm the independent studies by Vianna et al.[22] who used a smaller section of Love Canal for the study population and no local control. Since this is the first time a study on wildlife or stature of children has been used at a toxic waste site, these studies need to be replicated to determine whether they are generally useful. Each of these three objective measures showed a difference between Love Canal and control at less cost and in a more convincing manner than the subjective data on health problems as determined by interview.

REFERENCES

1. Legator, M., Ed. The Health Detective's Handbook. Johns Hopkins University Press, 1983.
2. Paigen, B. Controversy at Love Canal. The Ethical Dimensions of Scientific Conflict. Hastings Center Report, pp. 29–37, June 1982.
3. Levine, A. G. Love Canal—Science, Politics, and People. Lexington Books, Lexington, MA, 1982.
4. Paigen, B. and Highland, J. J. Methods for assessing health risks in populations living near hazardous waste sites. In: Highland, J. J., Ed., Hazardous Waste Disposal—Assessing the Problem, pp. 147–157. Ann Arbor Science, Ann Arbor, MI, 1982.
5. Battaglia, C., and Simmons, M. A. The low birth weight infant. In: Falkner, F. and Tanner, J. M., Eds. Human Growth, Vol. 2 Postnatal Growth. Plenum Press, New York, 1978.
6. Brandt, I. Growth dynamics of low-birth-weight infants with emphasis on the perinatal period. In: Falkner, F. and Tanner, J. M., Eds. Human Growth, Vol. 2 Postnatal Growth. Plenum Press, New York, 1978.

7. Schlesselman, J. J. Sample Size Requirements in Cohort and Case Control Studies of Disease. *Am. J. Epidemiol.* 99: 381–384, 1974.

8. Eveleth, P. B. and Tanner, J. M. Worldwide Variation in Human Growth. Cambridge University Press, Cambridge, MA, 1976.

9. Harrison, G. A., Weiner, J. S., and Tanner, J. M. et al. Human Biology. Oxford University Press, London, 1977.

10. Finkel, A. J. and Duel, W. C., Eds. Clinical Implications of Air Pollution Research. A.M.A. Publishing Services Group, Acton, MA, 1976.

11. Greulich, W. W., Crismon, C. S., Turner, M. A., Greulich, M. A., and Okumoto, Y. The Physical Growth and Development of Children Who Survived the Atomic Bombing of Hiroshima or Nagasaki. *J. Pediat.* 43: 421, 1953

12. Reynolds, E. L. Growth and Development of Hiroshima Children Exposed to the Atomic Bomb. Three Year Study (1951–1953). Atomic Bomb Casualty Commission, Technical Report 20–59, 1954.

13. Nehemias, J. V. Multivariate Analysis and the IBM 704 Computer Applied to the ABCC Data on Growth of Surviving Hiroshima Children. *Health Phys.* 8: 165, 1962.

14. Sutow, W. W., Conrad, R. A., and Griffith, K.M. Growth Status of Children Exposed to Fallout Radiation on the Marshall Islands. *Pediat.* 36: 5, 1965.

15. Burrow, N., Hamilton, H. B., and Hrubec, Z. Study of Adolescents Exposed In Utero to the Atomic Bomb, Nagasaki, Japan. *J. Am. Med. Assoc.* 192: 5, 1965.

16. Yoshimura, T. and Masato, I. Growth of School Children with Polychlorinated Biphenyl Poisoning or *Yusho*. *Environ. Res.* 17: 416–425, 1978.

17. LeQuesne, P. Clinical Expression of Neurotoxic Injury and Diagnostic Use of Electromyography. *Environ. Health Pers.* 26: 89–95, 1978.

18. Otto, D. Multidisciplinary Perspectives in Event-related Brain Potential Research. Proc. 4th Intern. Cong. on Event-Related Slow Potentials of the Brain. Environmental Protection Agency 600–9-77-043, p. 409, 1978.

19. Thomas, R. E. and Bellis, E. D. An Eye-lens Weight Curve for Determining Age in *Microtus Pennsylvanicus*. *J. Mammal.* 61: 561, 1980.

20. Rowley, M. H., Christian, J. J., Basu, D. K., Pawlikowski, M. A., and Paigen, B. Use of Small Mammals (Voles) to Assess a Hazardous Waste Site at Love Canal, Niagara Falls, New York. *Arch. Environ. Contam. and Toxicol.*, 12: 138–144, 1983.

21. Paigen, B. Health Hazards at Love Canal. Testimony Presented to the House Subcommittee on Oversight and Investigations, March 21, 1979.

22. Vianna, N. J., Polan, A. K., Regal, R., Kim, S., Haughie, G. F., and Michess, D. Adverse Pregnancy Outcomes in the Love Canal Area. Provisional draft released 1980. New York Department of Health, Albany, NY, 1979.

23. Paigen, B. Assessing the problem—Love Canal. In: Highland, J. J., Ed. Hazardous Waste Disposal—Assessing the Problem, pp. 14–19. Ann Arbor Science, Ann Arbor, MI, 1982.

CHAPTER 10

Integration of Government Resources in Assessment of Hazards

Robert D. Stephens

INTRODUCTION

Government agencies at various levels must, for a variety of reasons, assess the hazards posed by toxic wastes. These range from establishing proper operating standards and siting criteria for waste management facilities, to criteria for abandoned site cleanup. Resources to conduct these hazard assessments commonly exist in not one, but several public and private entities. There is, therefore, a necessity to combine and coordinate these resources to carry out the required assessments. Approaches to carrying out hazard assessment of different types are discussed. Models for intergovernmental cooperation are presented using examples of actual hazard assessments carried out at two hazardous waste facilities.

Public concern about hazardous wastes is today at a level unequaled by any other issue in the environmental movement. Much of the energy generated by this concern is directed toward government agencies to manage or eliminate hazardous wastes, to clean up abandoned or problem waste sites, and to remedy the environmental problems these sites have caused. Although there are times when government action on hazardous waste

J. B. Andelman and D. W. Underhill (Editors), *Health Effects from Hazardous Waste Sites*
© 1987 Lewis Publishers, Inc., Chelsea, Michigan – Printed in U.S.A.

problems is or would appear to be based upon political grounds, responsible action should be based upon an identification of the public health and environmental hazards which exist and through evaluation of the magnitude of the associated risks. Hazard evaluation and risk assessment of hazardous waste problems is, even in its simplest form, difficult. More typically, these tasks are extremely difficult and require extensive multidisciplinary resources. Given the necessity, however, of such activities, and the current limitations on technical staff, it is imperative that a process be established to integrate available resources.

There are two programmatic areas for which hazard evaluation and risk assessment are essential.* They are the siting and operational standards of hazardous waste generating, treatment, storage and disposal facilities, and the location, evaluation, and cleanup of problem or abandoned waste sites. The goals of these two related activities are similar, but the process and organization of the required investigation is different. For these reasons, evaluation of active sites and abandoned sites are discussed separately.

ACTIVE HAZARDOUS WASTE FACILITIES

States are currently involved in the process of granting final facility permits for all treatment, storage, and disposal facilities. The primary purpose of these final permits is to establish operational standards for hazardous waste facilities which result in minimum adverse public health and environmental impact. There is a wide variety of possible adverse impacts from a hazardous waste facility; some are aesthetic, and some are substantive impacts resulting from releases or potential releases of toxic substances to the environment. Analysis of the potential for environmental impact fundamentally relies on a knowledge of what substances will exist at the facility to be permitted and their behavior under conditions at the site. These two factors are most difficult to address at a hazardous waste land disposal facility. Hazardous wastes are in reality extremely complex mixtures of, commonly, large numbers of diverse chemical substances. Determination of the composition of wastes, in sufficient detail to allow for a reliable prediction of environmental impact, is almost never done. Behavior of chemical mixtures in environmental media is not completely understood. This is particularly true for the soil column in land disposal operations.

In addition to permitting existing facilities, state and local government officials must deal with the siting of new facilities. The fundamental siting questions to be answered are the hazards posed by the facility and risks experienced by the surrounding communities. The need for new facilities is

---

*In the context of this chapter, "Hazard" is meant as the intrinsic property of a substance, i.e., toxicity, explosivity, etc. Risk is related to a probability of adverse health impact based upon exposure, dose, and time.

clear if the country is to avoid a major waste management crisis. EPA and the National Solid Waste Management Association estimate that between 50 and 60 new sites will be needed over the next several years,[1] and there is reason to believe that the need for new facilities is even greater.[2] In most states, the siting decision process is more of a shared responsibility between state and local government than is the facility permit process. Siting is also more of a public process. Given the public interest in hazardous waste, a significantly greater burden is placed upon hazard evaluation and risk assessment during the public review process.

ORPHANED HAZARDOUS WASTE SITES

Programs designed to remedy or clean up old hazardous waste sites are in two phases in each of which a hazard and risk evaluation must be conducted. The first phase is site location and preliminary evaluation. The second phase is a comprehensive evaluation leading to remedial design, construction, and environmental damage assessment. Characteristics of each activity are discussed.

Preliminary Evaluation

The cleanup of problem waste sites begins with location and characterization of the sites. Most site location programs involve record searching and superficial physical survey. These processes commonly locate large numbers of potential problem sites. The first such comprehensive survey conducted in the United States disclosed approximately 60,000 potential sites.[3] The Abandoned Site Project in California surveyed 13 of the state's 58 counties and located 23,124 potential sites. Sites were located from a broad database using federal, state, and local government records, as well as industrial and public interest organizations.[4] Creation of such extensive lists of potential problem sites required a system of preliminary evaluation and priority ranking to determine which sites actually pose significant public health or environmental hazard. The large number of sites to be assessed precluded any intensive investigations. Evaluations were done primarily using existing data or data which could be obtained readily. No new hydrogeological data was obtained, and only a very limited amount of chemical analysis was conducted. The scheme used for this primary hazard evaluation is shown in Table 1.

In California, even this general information about sites is the province of several federal, state, and local governmental agencies. This was particularly true for much historical information possessed by field professionals in fish and game departments, water quality agencies, departments of health, and departments of public works. It is necessary on a county by county basis to form working groups with personnel from the proper agen-

Table 1. Prioritization Guide-Abandoned Site Project

1. Type of Waste
   a. Extremely hazardous or hazardous
   b. Unknown
   c. Nonhazardous

2. Location of Waste
   a. Waste visible
   b. Evidence of waste
   c. No indication of waste

3. Potential Groundwater Contamination
   a. Underlain by useable groundwater
   b. Underlain by unuseable groundwater
   c. No groundwater

4. Danger to Public Health
   a. Waste accessible to public, i.e., unsecured site or highly mobile waste
   b. Nearby development, secured site, low mobile waste
   c. Evidence of waste, unsecured site
   d. Waste visible, secured site, low mobile waste, rural location
   e. No indication of waste, secured site

cies to review and quality assure the historical data. Experience has shown that the quality of historical data relating to land use, business activity, and past dumping practices varies tremendously. Without review and quality assurance by persons having local knowledge, coupled with review by persons with chemical and engineering expertise, much historical data is of limited value and can be misleading. The key to organizing such intergovernmental working groups is to centralize project and organizational responsibility. The responsible agency acts as collector, organizer, and rapporteur of the data. This limits the required commitment of the contributory agencies to periodic working, data review sessions.

Comprehensive Site Evaluations

In-depth evaluations of problem waste sites serve a different purpose. Once a site has been determined to pose a significant public health or environmental hazard, extensive studies are required to:

1. Quantitate the magnitude of the toxics deposition
2. Delineate the limits of contamination
3. Characterize in detail the chemical substances present at the site
4. Assess the migration or migration potential of the specific substances at the site
5. Evaluate the impact of migrated or potentially migrated substances on target populations or environmental media
6. Evaluate the feasibility of various remedial strategies applicable to the chemical profile of the site and the existing environment

7. Evaluate the impact of the "cleaned" site on the current and future land use on and near the site

In contrast to the preliminary hazard evaluations discussed earlier, much of the data for the comprehensive site evaluations must be generated. Strategies and approaches to comprehensive site characterization have been discussed elsewhere.[5] Specific data needs for hazard evaluation are typically site specific; however, a factor common to all is the requirement for a significant amount of chemical, laboratory, engineering, environmental, and health expertise. Seldom is the level or breadth of expertise within one public or private entity sufficient. Commonly the agency responsible for conducting site cleanup and managing remedial fund has little of the required expertise. Such is the case with many state governments.

The key to successful comprehensive site evaluations is the ability to obtain staff and budgetary resource commitments from the affected federal, state, and local agencies. Although the relative resources of these three levels of governments vary from locale to locale, in our experience each level has a unique and necessary contribution. In addition to the actual technical resources which each may be able to contribute, as is discussed in some detail later, each has a unique perspective to problem site resolution. This perspective reflects the respective local, regional, or national constituency. The integration of these perspectives will be highlighted in examples given later.

An issue which always confronts a responsible governmental or private entity is the balance between the studies and hazard evaluation and the actual cleanup and remedial work. Significant competing pressures exist between the scientists' and engineers' desire to fully evaluate site impacts and characteristics leading to the best possible solution, and that of an affected community for rapid site cleanup. In California, the responsible agency, the Department of Health Services, has taken the position that remedial action on the basis of incomplete knowledge of site characteristics and behavior, particularly during remedial work, poses greater risk to the public health than nonaction. This position on several occasions has caused considerable displeasure and outright anger among public groups and local government. Problems caused by this policy have been significantly reduced by allowing for or insisting upon early involvement of local groups in problem resolution.

CASE EXAMPLES

Two examples are discussed which represent successful interagency programs to evaluate hazardous waste site impacts. The first is that of an operating landfill, and the second is that of an abandoned problem site. In both situations, resources were not available within any one federal, state, or local organization, and an interagency task force was assembled out of necessity.

Case 1. The BKK Landfill, West Covina, California

The BKK Company operates a large Class I landfill in the city of West Covina.* At this facility, liquid wastes are commingled with municipal solid waste and covered at the end of each day. Between the mid-1970s and 1981, process waste from a polyvinyl chloride plant was accepted at the facility. This waste contained up to 2% by weight of vinyl chloride monomer.

In October 1980, odor complaints from residents in the neighborhood surrounding the BKK landfill in West Covina prompted the South Coast Air Quality Management District (SCAQMD) to monitor for odorous organic compounds in the vicinity of the site. In May 1981, vinyl chloride was detected in the samples from this monitoring. A subsequent sampling confirmed the existence of vinyl chloride at concentrations exceeding the California Air Resources Board (ARB) Ambient Air Quality Standard of 0.01 ppm (24-hour average). The SCAQMD announced its findings in June 1981.

Because of these exceedences, the California Department of Health Services (DHS) immediately banned further disposal of wastes containing vinyl chloride at the BKK landfill. An evaluation of the vinyl chloride data by the DHS indicated that although the levels created no public health emergency in the surrounding population, they were of sufficient concern to warrant prompt efforts to reduce these emissions from the landfill. Also in May 1981, an interagency task force consisting of the SCAQMD, DHS, City of West Covina, Los Angeles County Health Department, State Solid Waste Management Board, and the California Regional Water Quality Control Board (Los Angeles Region) was formed to develop a program to reduce odor emissions from the site. This task force recommended a series of actions to expand the landfill's gas recovery system, and to improve the efficiency of the waste gas incineration system. As a result of these efforts, the frequency of exceedence of the vinyl chloride standard and the odor complaints decreased significantly through the first half of 1982. These data were presented at a public meeting in the city of West Covina on July 1, 1982.

Continued periodic exceedence of the vinyl chloride standard, as well as a concern over emission of other substances and citizen concerns regarding their exposure to volatile carcinogens from the site and the need to determine the overall effectiveness of the emission mitigation measures

---

*Under California law, Class I landfills can, with a few exceptions, accept the full range of hazardous wastes.

prompted a subgroup of the task force consisting of DHS, ARB, and SCAQMD to embark on an expanded monitoring program.

Purpose. This study has three purposes: (1) to evaluate the effectiveness of the mitigation measures already taken, and to determine whether or not further measures are necessary to control emissions from the site, (2) to determine if other volatile carcinogens besides vinyl chloride, known to have been disposed at the site in the past, are also present in the community at levels that would represent a public health concern, and (3) to provide a more comprehensive database for assessment of the health risk posed by emissions of carcinogens from the BKK landfill in West Covina.

Compounds Selected for Monitoring. Previous sampling of the landfill gas by SCAQMD, ARB, Euteck, Inc., and the University of Southern California indicated the presence at significant concentrations of suspected or known human carcinogens other than vinyl chloride. As with vinyl chlorides the presence of these compounds can be associated with the prior disposal of hazardous wastes at the site. Nine volatile compounds identified by these organizations were selected for monitoring based on their concentrations, carcinogenic potential, and the analytical capabilities of laboratories involved. These compounds are listed in Table 2 along with information on their uses, waste sources, and concentration in the landfill gas. DHS considers all nine compounds to be known or suspect human carcinogens.

Study Organization. The study program to evaluate health impacts from air emissions was divided into three parts. Each part had a lead agency while the interagency task force maintained general project management. Ambient air sampling was the responsibility of the air quality district with technical support from the state ARB. Analysis of collected samples was shared by the SCAQMD and ARB. Health effects assessment was conducted by the DHS. Review of final findings was carried out by the entire task force with representatives of the impacted community.

Several technical problems were encountered in both sample collection and analysis. The problems were a result of complex and uncertain meteorology, ambient air sampling for variety of substances simultaneously, and sampling and analysis at or near the analytical limits of detection. These difficulties were overcome by very careful quality assurance procedures and extensive interlaboratory comparisons.

The most difficult aspect of the study was utilization of the emissions and ambient air data to assess the health impacts on the contiguous community. Professional staff resources trained in the required environmental toxicological or risk assessment techniques were quite limited and existing in only one agency, the Department of Health Services. Table 3 lists the compounds

Table 2. Compounds Selected for Monitoring

Compound (Synonym)	Uses and Probable Waste Sources	Concentration Range in Landfill Gas (ppm)
chloroethene (vinyl chloride)	Manufacture of PVC and other co-polymers; organic synthesis; adhesives for plastics; refrigerant; manufacture of vinyl chloride also a significant waste source	83–12800[1]
1,1-dichloroethene (vinylidene chloride)	Copolymerized with vinyl chloride to manufacture types of saran; adhesives for synthetic fibers	N.D.[2]–1200[3]
trans-1,2-dichloroethene (acetylene dichloride)	General solvent for organic materials, dye extraction, perfumes, lacquers, thermo-plastics, organic synthesis	N.D.[2]–800[3]
trichloroethene (TCE)	Metal degreasing, extraction solvent for oils, fats, waxes; solvent dyeing; dry cleaning; refrigerant and heat exchange fluid; organic synthesis; fumigant; cleaning and drying electronic parts	N.D.–1000[3]
tetrachloroethene (Perc)	Dry cleaning solvent; vapor-degreasing solvent; drying agent for metals; heat exchange medium; manufacture of fluorocarbons	N.D.–1500[3]
1,2-dichloroethane (ethylene dichloride)	Manufacture of vinyl chloride; organic synthesis; anti-knock agent in gasoline; paint, varnish, and finish removers; metal degreasing; soaps and scouring compounds; wetting and penetrating agent; ore flotation	N.D.–5000[3]
trichloromethane (chloroform)	Manufacture of fluorocarbon refrigerants and propellants; dyes and drugs; general solvent; fumigant; insecticides	5.1[4] 0.038 (ambient air)[5]
benzene	Manufacture of styrene, phenol, synthetic detergents, cyclohexane for nylon, aniline DDT, various other insecticides, fumigants; solvent; paint remover; rubber cement; anti-knock agent in gasoline	10–2000[3]
chlorobenzene	Manufacture of chloronitrobenzene, DDT, aniline; intermediate and solvent for organic synthesis	N.D.–500[3]

[1]Analytical Research Laboratories, Inc. "Report for BKK/Stauffer Chemical." June 17, 1981.
[2]N.D. = Not detected.
[3]Euteck, Inc. "BKK Landfill Odor Study Final Report." Prepared for City of West Covina. February 27, 1981.
[4]South Coast Air Quality Management District. "Report on Vinyl Chloride Emissions from a Class I Landfill Operation." Source Test Report C–81–96 A & B. June 10, 1981.
[5]University of Southern California. "Investigation of Odorous and Volatile Compounds for BKK Class I Landfill Site in the City of West Covina." Prepared for BKK Corporation. July, 1981.

Table 3. Summary of Ambient Air Data (ppb)[1]

			Station			
	A	B	C	D	E	F
chloroethene (vinyl chloride)	7.1–7.3	4.5–5.5	2.0–3.0	3.8–4.1	2.0–3.0	2.0–3.0
tetrachloroethene (Perc)	2.1–3.7	1.8–2.9	1.4–2.4	2.3–3.0	1.5–2.1	1.5–3.4
trichloroethene (TCE)	0.8–1.0	0.8–1.8	0.2–0.3	1.6–1.7	0.6–0.8	0.8–1.2
1,1-dichloroethene (vinylidene chloride)	1.1–1.3	0.7–1.0	0.1–0.3	0.7–0.8	0.1–0.4	0.3–0.4
1,2-dichloroethane (ethylene dichloride)	1.3–3.0	0.8–2.8	0.4–0.7	0.8–2.8	0.7–0.9	0.4–1.1
trichloromethane (chloroform)	0.3–0.5	0.3–0.6	0.2–0.6	0.6–1.0	0.2	0.2
benzene[2]	4.8	4.6	3.6	4.6	3.0	3.0

[1]Each pair of values represents the means of the two laboratory data sets calculated for the entire study period. On those days when a compound was not detected, its concentration was assumed to be equal to the limit of detection.
[2]One set of data used. All of the second set were below the limit of detection.

and ranges of concentrations found in the monitoring program. The following list summarizes the findings of the health effects assessment.

1. Ambient levels of several substances considered to be suspect carcinogens were found to be elevated in the downwind vicinity of the landfill relative to a remote area considered to be background.
2. Levels of exposure to substances emitted at the landfill were 100- to 1000-fold below the threshold for acute noncarcinogenic effects.
3. Calculated cancer risk estimates placed increased accumulated individual risks at between 5 and 10 per 100,000.
4. Considering the total exposed population (ca. 8000) increased cancer rates are not to be expected. (Examination of cancer rates is possible through the Los Angeles Tumor Registry operated by the University of Southern California). No actual rate increases were noted.

Fundamentally, the study concluded that suspect carcinogens were emitted from this landfill, measurable environmental degradation had occurred, and a small but significant increase in individual risk to cancer could be estimated; however, no public health emergency existed which mandates drastic action such as community evacuation or site closure. This conclusion was extremely difficult to communicate to the community. A public meeting was called by the interagency task force to present the findings and conclusion of the study. The meeting was attended by approximately 500 persons. Response to the study conclusions was one of dismay and anger. Little success was achieved in communicating the meaning of low-level estimated risks to an emotionally charged, polarized community.

Case 2. The McColl Abandoned Waste Site, Fullerton, California

Between 1942 and 1946, several Southern California oil refineries used a then remote site for the disposal of spent and waste from the production of high octane fuels. The waste material was composed of heavy petroleum fractions and from 10 to 50% sulfuric acid. During the period from 1950 to 1960, attempts were made to cover and fill the acid petroleum sumps with drilling muds. The filling operation was never completed, the property was sold, and residential development began (early 1970s). Odor and health complaints began in 1976. Ultimately the state listed the McColl site as a highest priority waste site for Superfund cleanup. In late 1981, the state organized a task force called the "Participants Committee" to conduct a study which would assess the magnitude and character of hazards posed by the McColl site, and to provide sufficient data to allow a decision on the best remedial solution. The Participants Committee membership included:

California Department of Health Services
Orange County Health Department
State Water Resources Control Board

Air Resources Board
South Coast Air Quality Management District
Santa Ana Regional Water Quality Board
City of Fullerton
U.S. Environmental Protection Agency
Shell Oil
Texaco
Union Oil
McCaully Oil

This committee formed a subcommittee called the Technical Advisory Committee composed of technical experts in the requisite areas to assess:

Surface/groundwater
Air quality/odors
Health effects
Soils/waste materials

The Department of Health Services, as the responsible state agency, chaired both committees. An early decision was made that the required site assessment was beyond the staff resources of the participants and that external contracts were necessary. An unusual agreement was negotiated for shared funding of the studies between the state and several oil companies. Total cost of the initial studies was approximately $1 million.

It was clear to all participants that the findings of the study relative to environmental and health impacts affected future potential liability of the private participants. Protection against this fact influencing study conclusion was provided by state management of the study and commitment by public agencies of sufficient quality technical resources to track and understand the work. This strategy appears to have been successful.

The physical assessment of the site was conducted by contractor as overseen by the Participants Committee. This included geological, hydrogeological studies, waste characterization, and air studies. Health effects studies were done separately and conducted by the DHS. The study was in four parts: an adult health study, a pediatric health study, a reproductive outcome survey, and a pet health survey. The premise for the investigation was that airborne chemical emissions from the site might be causing adverse health effects. The specific objectives of the survey were to: (1) document the extent of the odor problem, (2) document the extent and severity of bothersome symptoms, (3) assess the patterns of medical care utilization to see if symptoms or illnesses severe enough to warrant medical care were in excess in the McColl neighborhood, (4) assess the incidence of malignant and benign tumors in the residents living around the site, (5) assess the incidence of untoward reproductive effects, and (6) assess the incidence of mortality and cancer among pets.

The epidemiologists from the State Department of Health Services recognize that the above objectives relate to events which have occurred in the few years which have elapsed since the construction of the homes surrounding the McColl site, and therefore the survey findings are not helpful in assessing human long-term effects which might conceivably result from living near the site. For this reason, a review of the toxicology of airborne substances was undertaken to assess possible long-term effects.

During the spring of 1981, a census was carried out in the neighborhoods south of the McColl site and in two control neighborhoods about five miles east of the site. The McColl neighborhood and the two control neighborhoods were similar in age and housing type. Questionnaires were sent to individuals in all of the households with instructions that adults fill in their own questionnaires and that an adult member of the household should provide the information on children under 16 years of age. Among adults enumerated in the census, 1024 responded to the survey which represents 82% of the adults in the McColl neighborhood and 69% of the adults in the control area. There were responses for 448 children which represents 73% and 64% of the children in the McColl and control neighborhoods, respectively. Information was obtained on about 70% of the pets owned by McColl neighborhood residents and about 60% of the pets in the control area.

The analyses of the survey data focused on comparing the responses of individuals in the control area to those of individuals living around the McColl site. For many analyses, the McColl neighborhood was further divided into five odor zones ranging from "no odor detected" to "odor often very strong." The odor zones were defined independently by researchers from TRC Environmental Consultants, Inc., who carried out an odor study in the spring of 1982. This further division of the McColl neighborhood made it possible to compare high and low odor zones within the community adjacent to the site.

Summary of the Major Findings

1. Both adults and children show an excess of respiratory symptoms, eye irritation and nausea. The dose-response relationship is clearer in the adults where there are larger numbers, but, nonetheless, there is considerable similarity in the symptom complaints of adults and children. The results are unlikely to have occurred by chance. Judging by the rates in the control area, there are anywhere from 50 to 100 more individuals complaining of these problems in the McColl area than would be expected.
2. Among adults (males and females) complaints of odor are much more common than in the control area. On the basis of the control neighborhoods rates, one would expect only 17 people to complain of odors in the

McColl neighborhood. In fact, there were 166 individuals out of 670 (25%) who said they could smell odors several times a week.

3. The proportion of children who had consulted a physician was similar in the two areas, but the number of consultations per child was greater in the McColl neighborhood. The reported reason for visits to a doctor were for conditions with the kind of symptom complaints reported in the questionnaire.

4. Based on parental reports, social and intellectual skills in children were not impaired as a result of living near the McColl site.

5. We were unable to detect any excess rates of cancer in adults or children or any excess rates of prematurity, malformations, miscarriages, or stillbirths.

6. Based on the reasons given for seeking medical care, we did not find that adults or children were experiencing any serious medical problems other than the symptoms.

7. About 26 more women than expected in the McColl area complained of developing menstrual problems since moving to their current address.

8. There was no difference in the mortality or cancer rates in the reports we received about household pets.

9. The population living around the McColl site numbers only about 1,000 children and adults. This means that only a very marked effect of the site could be demonstrated in an epidemiological study. This is true, both for a short-term study such as the present one, or even for a lifelong followup. The present study has demonstrated a marked difference in odor and symptom complaints and a moderate difference in number of doctor visits for children. We have not demonstrated differences in cancer, miscarriages, and the like, and would not be able to detect anything but very, very, marked differences, even in a lifetime followup. The chances are small that further study of this matter would provide definitive answers on cancer or reproductive effects.

These findings were presented to the McColl community with considerable success, in that they corresponded closely to undocumented ad hoc impressions of the health impacts of the site. In addition, there was a close correlation between the possible causes of the health symptoms and the chemical substances described by the physical chemical study of the site.

The findings of the health study were utilized by the interagency task force in the consideration of the variety of available remedial technologies. The various alternatives each have significant potential impact upon the contiguous community similar to those documented in the health study. Procedures for site cleanup were established to minimize these health effects, and health monitoring systems are to be established to identify quickly any such symptoms.

CONCLUSIONS

Integration of governmental resources for the assessment of hazards at waste facilities is desirable and even necessary in many cases. There is little disagreement on this point. Multiagency projects can however suffer from the classic "committee" syndrome where much time and effort is spent at accomplishing very little. Experience with these projects has shown that these problems can be avoided with proper study structure and management. Particularly important considerations include the following.

Structure of Interagency Group

It is critical that all the necessary technical and jurisdictional authorities be represented in the group. It is equally important that the role and responsibility of each be clearly delineated at the beginning.

Authority

The authority and purpose of the group should be clearly understood. Is the group advisory, scientific, or regulatory in nature? Relationship of participants, handling of data, and type of conclusions reached may vary significantly between a technical study and a regulatory action.

Project Management

Interagency groups, particularly larger ones, need a clear management structure with one, or at most two agencies as management. The managing agency should be the one with the primary authority over the facility or entity under study.

Budgeting

Studies of waste facilities are rarely budgeted items for several agencies. Such programs are typically described in governmental budgetary language as "bureaucratic overlap." Lack of appropriate budgets can spell failure for a study task force. Agency management needs to be able to sell interagency concepts in the budget process.

The requirement for and the act of utilizing an interagency group to a hazard assessment task complicates and commonly delays the final outcome. The political and public atmosphere existing today demands immediate answers to questions of hazards to the public. Considerable effort is required to defend and explain the necessity of conducting in-depth interagency studies.

ACKNOWLEDGMENT

The program elements described in this chapter have been created by various members of the California Department of Health Services. Principally among those have been Mr. Howard Hatayama, Hazardous Waste Management Branch; Drs. Raymond Neutra, Norman Gravitz, and James Stratton of the Epidemiological Studies Section; and Mr. Barton Simmons of the Hazardous Materials Laboratory.

REFERENCES

1. U.S. Environmental Protection Agency. Hazardous waste generation and commercial hazardous waste capacity — an assessment. SW–894, p. viii. U.S. EPA, Washington, DC, 1980.
2. Morell, D., Magorian, C. The hazardous waste crisis and the sitting imperative. In: Siting Hazardous Waste Facilities, p. 13. Ballinger Publishing Company, Cambridge, MA, 1982.
3. Therien, N. Report of abandoned site project. California Department of Health Services, 1983.
4. Castel, D., et al. Management of uncontrolled hazardous waste sites. U.S. Environmental Protection Agency, p. 275, 1980.
5. Stephens, R. D. Experimental design for wastesite investigations. In: William W. Lowrance, Ed. Assessment of Health Effects at Chemical Disposal Sites. The Rockefeller University, pp. 42–59, 1981.

SECTION V

Case Studies

CHAPTER 11

Assessment of Health Risks at Love Canal

Clark W. Heath, Jr.

INTRODUCTION

The toxic waste dump at Love Canal in Niagara Falls, New York, has been the focus of many environmental health studies over the past five years. As an early and highly visible example of what is now a major public health concern, the Love Canal situation and the studies that took place there deserve close attention. Future investigations of similar problems can benefit from understanding what kinds of studies were undertaken at Love Canal, what the results of those studies have been, and how those results can be interpreted.

On theoretical grounds, one can conceive of three general approaches to evaluating possible health risks in environmental exposure situations:

1. Measurement of the frequency of clinical illnesses potentially related to exposure
2. Measurement of subclinical effects which may predict the occurrence of clinical illness arising from exposure
3. Measurement of toxin levels in exposed persons and in the environment with prediction of illness risk based on current toxicologic knowledge

J. B. Andelman and D. W. Underhill (Editors), *Health Effects from Hazardous Waste Sites*
© 1987 Lewis Publishers, Inc., Chelsea, Michigan – Printed in U.S.A.

All three approaches were used in the Love Canal situation. This chapter reviews the results of these investigations and discusses their interpretation in relation to factors which limit the design and analysis of such studies.

CHRONOLOGY OF EVENTS

In the early 1890s, a community development project was begun in Niagara Falls by a developer named William Love. Central to his plans was a canal bringing water from the Niagara River around Niagara Falls to generate electric power. The project collapsed after only a portion of the canal was dug, a ditch 3000 feet long in the southeastern part of the city, a rural area at that time. Between 1942 and 1953 this ditch was used — primarily by the Hooker Chemical Company (HCC), which owned it — as a dump for chemical wastes arising mostly from the manufacture of pesticides. Records at HCC indicate that much of the waste material consisted of the pesticide Lindane (hexachlorocyclohexane) and its by-products (chlorobenzenes and other chlorinated hydrocarbons). In 1953, when the dump site was full, it was sealed with a clay cap. Over the next 10 to 15 years, the land surrounding the canal was developed as a residential neighborhood. Single-family homes were built along both sides of the canal, and an elementary school was constructed adjoining its central section. A street, a sewer line, and underground utility conduits were installed across the middle of the canal, damaging the clay cap.

By the mid-1970s, it became increasingly apparent that rainwater and melting snow had seeped through the broken cap into the dump itself. This seepage forced waste chemicals to the surface of the site and led to spread of chemicals through surface soil into basements of adjoining homes. In the summer of 1978, state and federal authorities purchased the first two rings of homes surrounding the canal (about 200 homes), relocated the residents, and began corrective drainage work. By the end of 1979, a system of drains linked to an onsite leachate treatment facility was completed, and the site was resealed with a covering of clay.

In the meantime, continued concern about possible health hazards at the site, together with the profound impact of the situation on property values and local economic stability, led in the early summer of 1980 to the evacuation of many homes in the general Love Canal area. Extensive environmental studies were performed at that time to determine the area's future habitability. Despite these efforts, the problem still remains unsettled in the face of pending litigation involving residents, HCC, and government agencies at local, state, and federal levels.

HEALTH RISK ASSESSMENTS

The first assessments of health risk at Love Canal were begun in the spring of 1978 by the New York State Department of Health (NYSDH).

Two kinds of investigation were carried out: first, environmental testing for toxic chemicals in soil, air, and water at the canal itself and in homes at various distances from the canal,[1] and second, a questionnaire survey of residents in the first two rings of homes around the canal (and expanded later to residents of the general area) regarding a wide range of acute and chronic health conditions which might be related to canal chemical exposures.[2] Although the latter survey had a very broad scope, it soon focused on reproductive health and in this context sought to review all pregnancy outcomes in persons living in the Love Canal area at any time since the 1950s. A further study was also carried out which examined rates of cancer occurrence in the Love Canal area.[3]

Later in 1978, and extending into 1979, additional environmental testing was carried out both by NYSDH and by the U. S. Environmental Protection Agency (EPA). At the same time, studies were begun by community groups working with scientists from the western New York area and with the Environmental Defense Fund (EDF). These studies initially sought information about the occurrence of many different illnesses in the area but later focused on measurements of childhood growth and development by comparing residents of the Love Canal area with residents of a nearby census tract.

In early 1980, a pilot study regarding possible chromosome aberrations in Love Canal residents was carried out with support from EPA.[4] This attempt to measure subclinical effects, together with a similar effort at the State University of New York at Buffalo (SUNYAB) with respect to measurements of nerve conduction velocity, raised sufficient concern about possible health risk that large-scale studies were begun in the summer of 1980 to reassess health patterns and environmental contamination in the Love Canal area. The latter study was performed by EPA and led to recommendations in July 1982 regarding future residential use of the area.[5] Plans for health studies were developed by the Centers for Disease Control (CDC) working with SUNYAB, but they were not undertaken for lack of funds. Instead, a cytogenetic study was conducted by CDC comparing residents of the Love Canal area with residents elsewhere in Niagara Falls.

MEASUREMENTS OF CLINICAL ILLNESS

Reproductive Health

The initial medical questionnaire survey performed by the NYDPH showed no differences in frequency for a wide range of acute and chronic illnesses when responses were compared between persons living in the first two rings of homes around the canal and persons living north of the canal.[2] Responses regarding reproductive abnormalities, however, did suggest some possible differences, and therefore that particular area of health was stud-

Table 1.   Reproductive Outcomes in Women Living in the Love Canal Area, Niagara Falls, New York, June 1978

Place of Residence		Total Pregnancies	Abortions		Total Live Births	Birth Defects		Low Birth Weight[b]	
			Number	Percent		Number	Percent	Number	Percent
Adjoining the Canal[c]		79	15	19.0	65	4	6.2	1	1.5
East in the Canal Area	Near Drainage Swales	108	25	23.1	85	10	12.0	13	15.7
	Not Near Drainage Swales	164	21	12.8	144	7	4.9	10	6.9
North in the Canal Area		125	11	8.8	110	8	7.3	3	2.7

[a]From NYSDH data.[2]
[b]Under 2500 grams.
[c]97th and 99th Streets.

ied in greater depth. The frequencies of three forms of reproductive abnormality were measured — spontaneous abortion, birth defects, and low birth weight (under 2500 grams) — for all women ever living in the Love Canal area. Reported abnormalities were checked with birth and medical records. Observed frequencies were compared between sections of the Love Canal area and with published frequencies from studies elsewhere. Since much attention in the fall of 1978 came to focus on the possibility that canal chemicals might have spread preferentially through natural drainage patterns in the area (swales), particular comparisons were made between persons whose homes were built on or near such swales and persons whose homes were relatively distant from swales.

The overall results for these several comparisons are summarized in Table 1 for women living in the area as of June 1978 and covering all reproductive events for those women for the duration of their residence in the Love Canal area. No *consistent* differences were seen between canal groups, whether comparison was made between swale and nonswale areas or between persons living in the first two rings and persons living to the north of the canal (somewhat more distant from the immediate contamination problem). The data are obviously limited by their small numbers. Particular attention was given to frequency of low birth weight since these particular data were ascertained from birth certificate forms and hence were free of possible later recall bias on the part of survey respondents. Frequency of low birth weight for births to women living near swales did appear unusually high (13 of 85 live births, 15.7%). The interpretation of this isolated observation is unclear. Although the entire data set has yet to be fully analyzed, initial review suggested that abortion frequencies may have been

Table 2. Incidence of Selected Cancers by Sex, Love Canal Census Tract, 1966–1977
(Adapted from Janerich et al.[3])

	Number of Male Cases		O/E Ratio[b]	Number of Female Cases		O/E Ratio
	Observed	Expected[a]		Observed	Expected[a]	
Digestive organs	13	15.7 (8–24)	0.8	15	13.8 (7–22)	1.1
Respiratory system	25	15.0 (8–23)	1.7	9	4.6 (1–8)	2.0
Genital organs	13	7.9 (3–14)	1.6	8	12.5 (6–20)	0.6
Urinary system	7	6.0 (2–11)	1.2	1	2.4 (0.5)	0.4
Lymphoma, leukemia, liver cancer	3	6.3 (2–12)	0.5	6	4.7 (1–9)	1.3
All sites	71	60.8 (46–77)	1.2	71	65.4 (50–82)	1.1

[a]95% confidence intervals in parenthesis.
[b]Observed/expected ratio standardized for age.

increased for women living adjacent to the canal in the 1960s but not later. While such a pattern might suggest an association with dump exposure at a time when dumping was active, there remains no indication for such an effect in later years when dump containment became increasingly compromised.

Cancer

Since much of the waste material in the Love Canal represented organic chemicals with oncogenic potential, it was appropriate to examine patterns of cancer occurrence in the Love Canal area.[3] This was facilitated by the existence of a statewide tumor registry which enabled site-specific incidence in the Love Canal census tract to be compared with incidence elsewhere in Niagara Falls and New York state. This was done with particular attention to malignancies of blood cells, lymphoid cells, and liver, and tissue sites which might be at particular risk of malignant transformation from exposure to chemicals of the sort present at Love Canal. Data from this study are summarized in Table 2. Information was examined separately for males and females since occupational exposures might be expected to obscure exposures which occurred at places of residence. Aside from increases in incidence of respiratory cancer in both sexes, no significant changes were seen for cancers occurring in the 12-year time period, 1966 to 1977. The increases in lung cancer were considered more likely due to local patterns of cigarette smoking than to exposures at the dump site. Cases of lung cancer did not cluster geographically in relation to the canal itself, and increased observed/expected ratios as large or larger than those seen in the Love Canal census tract were found for either males or females in seven of 24 other census tracts in Niagara Falls. Leukemia, lymphoma, and liver cancer occurrence did not show any tendency to have increased over time since 1955 in the Love Canal tract.

Three methodologic problems make interpretation of these data a difficult matter.[6] First, the population studied is relatively small, considering the relative rarity of cancer at particular sites over short periods of time. Since the numbers of cases are therefore small, analysis is limited. Second, the area covered by the Love Canal census tract is relatively large compared to the area of the canal itself with the first two rings of adjoining homes. This means that only a small proportion of the census tract population can realistically be considered as living truly close to the canal. In studying the entire census tract as a unit, therefore, one is necessarily including in denominators used for calculation of rates a large proportion of individuals whose exposure to the canal is relatively low, thus diluting rates in "exposed" persons. Third, since the latent period for cancer development after chemical exposure appears to involve at least 10 to 15 years (more perhaps, if dose levels are low), it seems unlikely that increased incidence would yet have appeared, particularly if exposure from canal leakage was greatest in the 1970s. Under such conditions, the question would remain unresolved until a decade or two hence, assuming of course that exposure can be defined and that sufficient persons were exposed to permit an increased risk to be detected at that time.

MEASUREMENTS OF SUBCLINICAL ABNORMALITIES

A second general approach used in the Love Canal situation for detecting possible health effects was the indirect method of measuring certain subclinical abnormalities with intent to relate findings to eventual risk of illness. Work was carried out with respect to frequency of chromosomal aberrations and to measurements of nerve conduction velocity. In the latter instance, a pilot investigation carried out at SUNYAB found no significant differences in conduction velocity between two small groups of subjects, one from the Love Canal area and one from a control census tract to the west. Further neurologic testing was planned to enlarge the sample through the joint CDC-SUNYAB study, but that study did not materialize.

With respect to cytogenetics, frequencies of chromosome breakage in 36 volunteers from the Love Canal area were measured in blood specimens obtained in January 1980. Work was performed by the Biogenics Corporation as a pilot study under contract with EPA.[4] Among 7102 cells scored, 78 contained one or more chromosome aberrations (breaks, gaps, acentric fragments). Although no concurrent controls were used and specimens were not scored blindly in the laboratory, this level of breakage frequency (1.0%) did not differ substantially from levels observed by the same laboratory in other populations studied before. Concern was voiced, however, about the presence of large supernumerary acentric fragments in some specimens. Although the exact significance of these fragments was unclear, reports made to the persons in whom the cytogenetic testing was done suggested a possible relationship to future risk of cancer and genetic disorders.[7-8] Concern raised by this interpretation of these preliminary observations[9-11] led to plans for a more complete cytogenetic analysis among Love Canal residents as part of the larger CDC-SUNYAB study which was planned during the latter part of 1980. When that study proved impossible to fund, a smaller cytogenetic investigation was conducted by CDC beginning in the fall of 1981. Work encompassed specimens from past residents of first ring homes where NYSDH testing in 1978 showed elevated levels of chemicals likely to have arisen from the canal, and efforts were made to repeat cytogenetic analyses on persons studied by Biogenics in 1980. Matched control specimens from another census tract in Niagara Falls were interspersed with Love Canal specimens for blind laboratory analysis. Specimens have been tested for frequency of chromosome aberrations and frequency of sister chromatid exchanges.

The interpretation of subclinical measurements such as cytogenetic observations is subject to the same limitations as clinical studies: uncertainty about the actual exposure status of persons tested, limitations in sample size, and considerations of latency. In addition, one must evaluate how long subclinical changes may persist, and hence be detectable, after biologic damage takes place, and to what extent such subclinical abnormalities predict the development of later clinical illness. In the case of cytogenetic abnormalities, it is by no means certain that increased frequency of chromosomal aberrations is linked to increased disease risk, however logical such an association may seem a priori.[12] To date, no evidence of such a link has yet appeared in the studies conducted to follow the health status of Japanese survivors of atomic bomb irradiation. Very little evidence has yet been developed regarding populations with chemically induced chromosome damage.

MEASUREMENTS OF ENVIRONMENTAL TOXINS

In each clinical or subclinical study undertaken at Love Canal, a major difficulty for study design and interpretation has been the lack of precise evidence defining exposure in specific individuals. The chemicals of concern are largely of a nonpersistent sort, meaning that one does not have the option of identifying degree of exposure by measurement of chemicals persisting in tissue or in fat storage. Tests for a battery of chemicals were conducted by EPA on sera from the 36 residents in whom chromosome studies were done, but these tests merely showed expected trace background levels. Chemical testing is, therefore, a matter of determining current levels of chemicals in the Love Canal environment and basing current and future health risk estimates on those values. Determination of past levels of chemicals in the area is clearly a matter of indirect extrapolation from current findings and from whatever records and historical information exist.

In the Love Canal situation, the earliest tests for chemical levels in the environment were begun in the spring of 1978 by the NYSDH.[1] This testing focused on samples of air, water, and soil obtained directly on the canal and from homes and yards directly abutting the canal. Testing was later extended to homes more distant from the canal, and particular attention was given to locations near the original swale drainage routes in the area. A wide range of chemicals, organic and inorganic, were tested, and measurements were also made to detect levels of radioactivity. From this large set of data, it appeared that, before evacuation of the first two rings of homes and before remedial drainage construction work began, increased levels of chemicals related to the canal were seen only on the canal site itself, in storm sewers and creeks draining the area and in certain homes in the first two rings, particularly at the south end of the canal. These values were subsequently used for selecting households in which to study levels of chromosome aberrations.

A second systematic program of environmental testing at the Love Canal was undertaken by EPA in the summer of 1980 after remedial construction was completed.[5] The purpose of this program was to determine degree of environmental contamination at that point in time as a basis for forming recommendations regarding future use of the area. A large number of chemicals were measured in a large number of environmental samples from the Love Canal area and from other parts of Niagara Falls and vicinity for comparison purposes. Samples of soil, air, and water were tested, and results showed patterns virtually the same as what NYSDH had observed in its earlier tests. Elevated levels of chemicals likely to have arisen from the canal were seen consistently only in certain parts of the inner two rings of homes and in storm sewer samples. Review of these data by the Department of Health and Human Services, with evaluation of technical methodologies by the National Bureau of Standards, led to federal recommendations in July 1982 that the general area surrounding the Love Canal was safe for human residence outside the canal itself and the two rings of homes surrounding it. It was also recommended that the storm sewers and their drainage tracts be cleaned and that special plans be made for perpetual maintenance of the clay cap covering the site.

DISCUSSION

If the results of environmental testing on and around the canal in 1978 can be taken as representative of levels of contamination experienced in the community during previous years (and it seems unlikely that contamination had diminished), exposure would seem to have been very small except on the site itself and at certain first and second ring homesites. Unfortunately, past levels of exposure and cumulative dose can only be inferred since no biologically persistent chemical markers were available for measurement. In any case, measurement of exposure and dose remains the single most important element in assessing health risk. Inability to make such measurements with precision for individual residents of the area has been perhaps the most telling handicap for developing productive epidemiologic observations.

Epidemiologic studies in the area, of course, were hampered by several other methodologic difficulties, as discussed previously. None of these difficulties was unique to the Love Canal setting since they are confronted whenever epidemiologic work is undertaken. They include, again, issues of latency where one deals with chronic and delayed illnesses like cancer, the problem of sample size in relation to frequency of disease and risk factor occurrence, and the fact that most clinical endpoints for disease caused by environmental factors are quite nonspecific and do not often permit one to conclude, on clinical grounds alone, that a particular case of disease resulted from a particular exposure.

Even with these limitations, however, it can be said from current epidemiologic data available at Love Canal that no striking increases in illness occurrence have thus far appeared in association with living near the Canal. This does not mean that such occurrences might not yet appear or that some canal-related illness may not have occurred at frequency levels not detectable by the studies performed. But the observations do correspond with the results of environmental testing which show a lack of substantial contamination outside the first two rings. None of these negative observations, of course, at all diminishes the need to maintain the security of the dump site, although one would hope that this could be done while restoring the surrounding neighborhood to normal activity. In the meantime, scientific attention needs to continue to focus particularly on methods of applying subclinical observations to environmental health risk assessment, including ways in which individual exposure levels can be objectively measured or inferred. The relatively small size of populations living near particular dump sites (the Love Canal situation being one with a relatively large population) calls for methods of investigation with greater power than measurement of relatively infrequent and etiologically nonspecific clinical illness.

REFERENCES

1. Kim, C. S., Narang, R., Richards, A., et al. Love Canal: Chemical contamination and migration. In: Proceedings of the National Conference on Management of Uncontrolled Hazardous Waste Sites, Environmental Protection Agency, pp 212-9, 1980.
2. Vianna, N. J. Adverse pregnancy outcome - potential end points of human toxicity in the Love Canal. Preliminary results. In: Porter, I. H., and Hook, E. B., Eds. Human Embryonic and Fetal Death. New York, Academic Press, pp 165-8, 1980.

3. Janerich, D. T., Burnett, W. S., Feck, G., et al. Cancer incidence in the Love Canal area. *Science* 212:1404–7, 1981.
4. Picciano, D. Pilot cytogenetic study of the residents living near Love Canal, a hazardous waste site. Mammalian Chromosome Newsletter 21:86-93, 1980.
5. Environmental Protection Agency. Environmental monitoring at Love Canal. 3 volumes. Washington, U.S. Government Printing Office. Document No. EPA-600/4–82–030a, 1982.
6. Heath, C. W., Jr. Field epidemiologic studies of populations exposed to waste dumps. *Environ. Health Pers.* 48:3–7, 1983.
7. Kolata, G. B. News and comment. *Science* 208:1239–42, 1980.
8. Culliton, B. J. News and comment. *Science* 209:1002–3, 1980.
9. Shaw, M. W. Letter to editor. *Science* 209:751–2, 1980.
10. Gage, S. J. Letter to editor. *Science* 209:752–4, 1980.
11. Picciano, D. Letter to editor. *Science* 209:754–6, 1980.
12. Bloom, A. D., Ed. Guidelines for Studies of Human Populations Exposed to Mutagenic and Reproductive Hazards. New York, March of Dimes Birth Defects Foundation, 1981.

CHAPTER 12

Adverse Health Effects at a Tennessee Hazardous Waste Disposal Site

Robert H. Harris, Joseph V. Rodricks, C. Scott Clark, and
Stavros S. Papadopulos

INTRODUCTION

In the early 1960s, pesticides leaking from a Memphis landfill were alleged to be responsible for major fish kills along the Mississippi River south of Memphis.[1] Subsequently, Velsicol Chemical Corporation, a major contributor to that landfill, transferred its hazardous waste disposal operations to Toone-Teague Road in rural Hardeman County, some sixty miles east of Memphis. Using what Velsicol termed the "time-honored practice [of] shallow burial of toxic wastes," over 300,000 55-gallon drums of solid and liquid pesticide production wastes were buried in shallow trenches at the site between 1964 and 1972, when the State ordered the dumping stopped.[2]

Hydrogeological studies in 1967 concluded that groundwater was flowing in a direction away from the local private drinking water wells along Toone-Teague Road.[3] However, by 1977 local residents were complaining about the taste and odor in their well water and reporting an unusually high incidence of health symptoms. These included skin and eye irritation; weak-

J. B. Andelman and D. W. Underhill (Editors), *Health Effects from Hazardous Waste Sites*
© 1987 Lewis Publishers, Inc., Chelsea, Michigan—Printed in U.S.A.

nesses in the upper and lower extremities; upper respiratory infection; shortness of breath; and severe gastrointestinal symptoms, including nausea, diarrhea, and abdominal cramping.[4] Unfortunately, the 1967 hydrogeological studies incorrectly predicted the direction of groundwater movement—pesticide wastes had slowly been seeping into groundwater and by 1978 had contaminated drinking water wells with extraordinarily high concentrations of a number of toxic and known carcinogenic chemicals (e.g., carbon tetrachloride, chloroform, and tetrachloroethylene) in addition to a number of other chemicals that may represent a chronic health threat.

After a brief discussion of the history of landfill operations at this site, the nature of groundwater contamination in the area, and preliminary health surveys of residents in the vicinity of the landfill, a risk estimate will be presented based on animal toxicity data. Exposure estimates are based on a groundwater model that predicts retrospective exposures from 1978, when domestic wells were abandoned, and from measurements and estimates of indoor air contamination from outgassing of the volatile organic chemicals identified in the groundwater.

HISTORY OF LANDFILL OPERATIONS

Velsicol's Hardeman County landfill is located on a 242-acre farm on Toone-Teague Road (Figure 1). Wastes were transported from Memphis primarily in drums and fiber cartons and were deposited in 15-foot wide and 12-foot deep trenches excavated along ridges in the predominantly rolling terrain of Hardeman County. The trenches were unlined, and bulldozers were used to crush the waste containers and cover the trenches with native soil. The wastes included the residues of manufacturing heptachlor, dieldrin, endrin, and heptachlor epoxide. In addition to these pesticides, the wastes were identified to contain chlordane, isodrin, hexachlorocyclopentadiene, hexachlorobicycloheptadiene, carbon tetrachloride, and chloroform.[2]

Dumping operations in Hardeman County commenced in 1964. In 1965, a Memphis county engineer became concerned that dumping operations in Hardeman County might be endangering the artesian aquifer supplying drinking water to Memphis.[1] Alarmed, the State of Tennessee requested that the U. S. Geological Survey (USGS) conduct a hydrogeological investigation of the area. In 1967, the USGS concluded that, ". . . neighborhood wells are . . . either upgradient or perpendicular to the gradient. There is, therefore, no possibility for any existing water table wells to produce potentially contaminated water. . . ."[3] The USGS also concluded that there was no possibility that the 200-foot deep artesian aquifer would become contaminated. However, surface water and soil down to the water table under the landfill site were shown to be contaminated with chlorinated hydrocarbons.

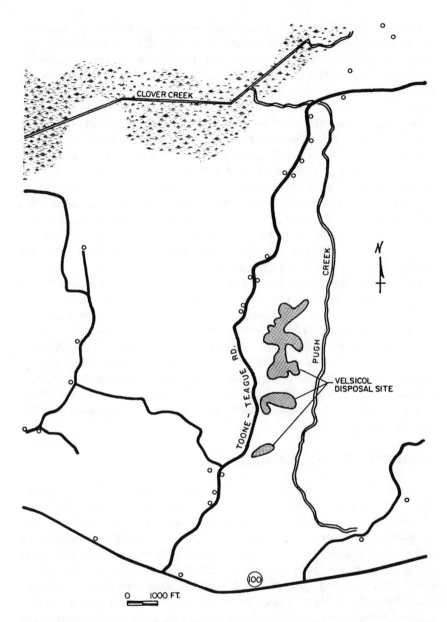

Figure 1. Map of region showing locations of Velsicol's disposal site and domestic wells in 1978.

Encouraged by the USGS report that the domestic wells in the area were not in danger, Velsicol expanded the 20-acre landfill both northerly and southerly to more than 40 acres over the following several months. Although local residents reportedly complained of air pollution from the dumping operations, there were no complaints of tainted well water at that time.

In 1970, a USGS hydrogeologist who did not participate in the 1967 study concluded that, notwithstanding the 1967 report, Velsicol's landfill did threaten local domestic drinking water wells primarily because Velsicol had expanded its dumping operations and more houses with domestic wells had been constructed near the dump site. After considerable debate among company, state, and local officials, in 1972 the State of Tennessee ordered Velsicol to close the landfill except for "nontoxic" wastes. All dumping operations were reported to have ceased by 1975, and in August, 1980, a clay cap was constructed over the landfill and seeded in grass to reduce further infiltration of water through the waste materials.

Following congressional hearings in October, 1978, and a second USGS hydrogeological report issued that same year, Velsicol admitted that the landfill was responsible for contaminating local domestic wells. However, because of persistent taste and odor problems, a number of residents had stopped using their well water for potable uses by the fall of 1977. A larger group of residents ceased potable uses in May, 1978, after warnings were issued by the county health department. By August, 1978, almost all local residents had ceased potable use, and by January, 1979, all uses of the contaminated well water had ceased and alternate drinking water supplies were provided.[4]

GROUNDWATER CONTAMINATION

The first hydrogeological investigation of Velsicol's Hardeman County dump site area was conducted by the U.S. Geological Survey.[3] The study indicated that the disposal site is underlain by a sequence of nearly horizontal strata that are composed of deposits of sand, silt, and clay (Figure 2). To a depth of about 100 feet, these strata consist predominantly of sand deposits interbedded with thin deposits of silt and clay. Between a depth of about 100 to 200 feet, beds of clay are predominant. The upper 250 feet of strata beneath the disposal site contain three distinct groundwater zones: a perched water zone, a water table aquifer, and an artesian aquifer. The water table aquifer, which supplies local domestic wells, occurs at a depth of about 90 feet below the surface of the disposal site and 60 feet below the base of the perched water zone. The 1967 USGS report concluded that "water entering or passing through the water table aquifer from potentially contaminated sources will tend to move northeasterly toward Pugh Creek." However, the water level elevations presented in the report indicate that

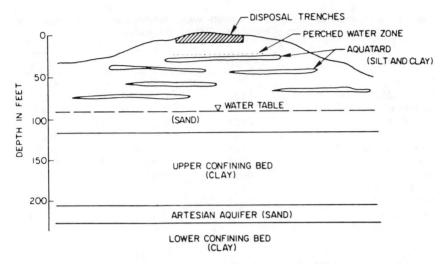

Figure 2. Schematic of geologic cross-section of Velsicol's disposal site in Hardeman County.

flow in the water table aquifer was predominantly toward the north or northwest, in the direction of local domestic wells (see Figure 1). Therefore, it is unclear which data Rima et al.[3] used to conclude that flow in the water-table aquifer was predominantly toward Pugh Creek, and away from local domestic wells.

Although the 1967 USGS report incorrectly determined the direction of flow in the water table aquifer, the report did clearly demonstrate that pesticide wastes had contaminated portions of both the surface and subsurface environment at the site. Evidence of contamination of the surface environment was found in samples of the surface soil at the disposal site, in samples of sediment washoff from the disposal site, and in samples of the bed load in Pugh Creek as far as 1.5 miles downstream from the site. In the subsurface, the buried wastes were shown to have produced a zone of contamination directly beneath and peripheral to the disposal pits.

Between 1967 and 1972, the area of the disposal was enlarged both northerly and southerly from approximately 20 acres to over 40 acres. Because of these changes and because questions arose as to the direction of leachate migration from the disposal area, the Tennessee Department of Public Health and the USGS began a cooperative study in 1976[2] to reexamine the site and review the conclusions of the 1967 study. This study indicated that contamination still existed in surface water, in sediments from Pugh Creek, in the shallow perched water table, in deeper perched water zones, and in the water table aquifer. Groundwater movement in the water table aquifer was predominantly toward the north and northwest, in the direction of

Table 1.   Contaminants Detected[a] in Private Wells Serving Exposed Residents in
Toone-Teague Area of Hardeman County, Tennessee[4]

Compound	NP[b]/NT[c]	Range (μg/L)	Median (μg/L)
benzene	7/7	5–15	12
carbon tetrachloride	15/15	61–18,700	1,500
chlordane	5/24	Trace–0.81	Trace
chlorobenzene	23/25	Trace–41	5.0
chloroform	14/15	2.1–1,890	140
hexachlorobutadiene	22/28	Trace–2.53	0.15
hexachloroethane	19/31	Trace–4.6	0.26
hexachlorobicyclo- heptadiene (HEX-BCH)	24/31	Trace–2.2	0.05
methylene chloride	11/11	1.5–160	45
naphthalene	13/13	Trace–6.7	ND[d]
tetrachloroethylene	27/38	Trace–2,405	3.5
toluene	14/24	0.1–52	0.6
xylenes	2/3	0.07–1.6	0.07

[a]U.S. Environmental Protection Agency, Region IV. March 9, 1979. *Summary of USEPA and State of
Tennessee Chemical Analysis*, Atlanta, Georgia.
[b]NP = Number of samples with detectable amounts of the substance tested
[c]NT = Total number of samples tested.
[d]ND = Not detected.

domestic wells. Although it was hypothesized that heavier-than-water liquid
wastes could be migrating away from the site along the surface of the upper
confining bed (Figure 2), this could not be determined with any degree of
certainty by the study.

To supplement the two USGS studies, Velsicol retained the consulting
firm of Geraghty and Miller, Inc., in July, 1978, to undertake an assessment
of groundwater conditions at the site. Geraghty and Miller Inc.[5] generally
confirmed the direction of groundwater flow reported by the USGS in 1978.
In addition, Geraghty and Miller, Inc.,[6] presented data showing high levels
of carbon tetrachloride and chloroform in some local domestic water wells.
Contaminant concentrations were shown to be variable over time, with
concentrations of carbon tetrachloride ranging up to 13,400 ppb in one
domestic well sampled in November, 1978.

Extensive groundwater sampling has been conducted by the U. S. Environmental Protection Agency (EPA) and the State of Tennessee since 1978.
Contaminants detected in private wells serving residents in the Toone-
Teague area in 1978 are presented in Table 1. Concentrations of carbon
tetrachloride, the contaminant found in the highest concentration, range up
to 18,700 ppb in one private well approximately 1800 feet north of the
northern boundary of the landfill. As is usually found with groundwater
contaminated from hazardous waste landfills, the concentrations of contaminants fluctuated over an approximate tenfold range during the year,

although water quality data compiled subsequently[7] have indicated increasing concentrations of most contaminants with time. For example, in one domestic well approximately 1500 feet north of the landfill, the concentration of carbon tetrachloride from November, 1978, through November, 1979, fluctuated between 9820 ppb to 20,000 ppb; from May, 1981, through June, 1982, the concentration of carbon tetrachloride ranged from 18,000 ppb to 164,000 ppb.

Most residents had stopped using the well water for potable supply in 1978 when the first groundwater samples were taken for organic chemical analyses. However, exposures to volatile organic contaminants (e.g., carbon tetrachloride) resulting from the outgassing of the chemicals into the indoor environment, presumably continued until all uses ceased in early 1979. In an effort to predict the concentrations of contaminants in the well water prior to 1978, S.S. Papadopulos & Associates, Inc.,[8] developed a model of groundwater flow and contaminant transport within the water table aquifer. The model was based on the hydrogeological studies of the USGS and Geraghty and Miller, Inc., and on the groundwater monitoring data of the U.S. EPA, the State of Tennessee, and AWARE, Inc.[9]

The model used for this evaluation[10] incorporates into a mathematical framework the physics of groundwater flow and the transport of chemical constituents dissolved in the groundwater. Model parameters that describe the physical characteristics of the aquifer system were obtained from the hydrogeologic studies of the site and its vicinity, and from a process of model calibration. The calibration process involved adjustments to some of the physical parameters to reproduce satisfactorily the observed groundwater level and contaminant concentrations. After calibration, the model was used to reconstruct the past history of the contamination in the vicinity of the dump site and to forecast potential future conditions.

The predicted average concentration of carbon tetrachloride, prior to groundwater sampling in 1978, in domestic wells located about 1500 feet north of the site boundary is shown in Figure 3.

HEALTH SURVEYS

During the fall of 1977, residents complained of unusual and unpleasant taste and odor in their drinking water and reported a wide range of symptoms including skin and eye irritation; weakness in the upper and lower extremities; upper respiratory infection; shortness of breath; and severe gastrointestinal symptoms, including nausea, diarrhea, and abdominal cramping.[4] Coincident with the EPA's warning in November, 1978, that residents should stop all uses of well water, Clark et al.[4] initiated a limited health survey primarily to ascertain whether exposure to the water was associated with liver dysfunction.

The population studied was divided into three groups: 61 individuals

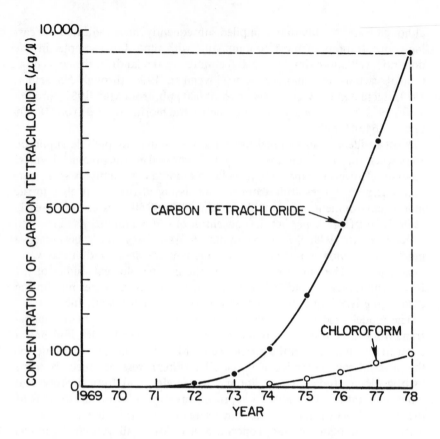

Figure 3. Model estimate of yearly average carbon tetrachloride and chloroform concentrations with time in a domestic well approximately 1500 feet north of the northern boundary of Velsicol's disposal site in Hardeman County.

from 16 households along Toone-Teague Road who used water with carbon tetrachloride concentrations exceeding 150 µg/L; an intermediate exposed group of 33 individuals exposed to carbon tetrachloride in drinking water at concentrations less than 45 µg/L; and an unexposed control population of 57 individuals recruited from civic organizations and church groups elsewhere in Hardeman and adjacent counties. The survey utilized a health questionnaire, a clinical examination, and biochemical screening. The latter included analyses of serum for liver and kidney function parameters, determination of hepatitis A and B serology, and bile acid determinations on fasting and postprandial serum and urine specimens.

The initial hepatic profile testing revealed elevated concentrations of the serum enzymes, alkaline phosphatase and serum glutamic oxaloacetic transaminase (SGOT), in residents who used the water. During followup examinations of these same persons two months later (January, 1979), these values were significantly reduced, as were postprandial serum concentrations of the bile acid cholylglycine.

In the 12-month period prior to the study, 11 Hardeman County residents were hospitalized. Complaints included gastrointestinal abnormalities, respiratory difficulties, neurological symptoms, muscular complaints, and high fevers. Other individuals reported symptoms such as dizziness, headache, skin rashes, nausea and vomiting, numbness of limbs, and menstrual irregularities. The only birth was a low birth weight infant with gastroschisis and eventration of intestines. Seven adults and five children reported vision problems. Four persons in the intermediate-exposed group reported hospitalizations: two for heart problems, one for breast surgery, and one for pneumonia. In the control group, there were five hospitalizations: one each for finger surgery, kidney stones, hip surgery, and splenectomy, and the other for undiagnosed abdominal pain.

Physical examinations were conducted on 118 individuals: 48 in the exposed group, 24 in the intermediate-exposed group, and 46 controls. During physical examination, seven individuals (six exposed and one intermediate) had borderline liver enlargement (slight hepatomegaly). A significant difference between the groups was found ($p = 0.034$) using the Pearson chi-square test. No kidney, skin, or eye abnormalities were detected.

In a subsequent and independent survey of residents exposed to contaminated groundwater near the Velsicol site, Rhamy[11] completed health histories and physical examinations on 112 persons living or formerly living within three miles of the landfill. Analyses of the data from these individuals were confined to 102 persons. Those excluded from the analyses were those who appeared not to have been exposed to the contaminated well water. Each symptom was asked six different times during the examination, and the symptoms were segregated as to those during development, prior to 1964, those from 1964 to 1972, and those from 1973 to 1982. The results of the interviews for symptoms reported from 1973 to the time of the survey (March, 1982) are presented in Table 2. The physical examinations on the 102 individuals revealed the following: diastolic hypertension, 28%; increased liver size, 8%; borderline hepatomegaly, 2%; optic atrophy, 7%; borderline optic atrophy, 4%; peripheral neurological changes, 10%; borderline peripheral neurological changes, 8%; and significant eye problems, 32%.

Although Rhamy did not include a control population in this study, he concluded that the extraordinarily high incidence of abnormal physical symptoms and clinically observed abnormalities were most likely related to the chemical exposures from the contaminated groundwater. Of particular note was the high incidence of headaches, often associated with showering when the odor in the air was described as "pungent." Rhamy also considered the high incidence of eye problems, particularly optic atrophy, to be note-

Table 2.   Symptoms Reported by Individuals Exposed to Contaminated Groundwater in Hardeman County.[4]

Organ	Symptom	Percent Reporting Frequent Symptom Since 1973, but not Before 1973
Head	Headache	88
	Faintness	55
	Dizziness	55
Eyes	Decrease in vision	26
	Burning of the eyes	69
	Bloodshot eyes	56
	Photophobia	53
Ears	Ringing in the ears	35
Nose	Burning in the nose	52
Throat	Soreness of the throat	48
	Burning of the throat	64
Mouth	Loss of taste	23
Heart and Lungs	Cough	26
	Shortness of breath	52
GI Tract	Nausea	52
	Vomiting	38
	Diarrhea	16
Urinary	Urgency and frequency	34
Skin	Skin rash	40
CNS	Tingling and paresthesias	51
	Loss of balance	21
	Muscular weakness	59
General	Lassitude	94

worthy, particularly since optic atrophy has been associated with occupational exposures to carbon tetrachloride.[12]

RISK ESTIMATES

Despite the inconclusive nature of the health surveys conducted by Clark et al.[4] and Rhamy,[11] it would appear that the evidence at least strongly suggests that contaminated groundwater along Toone-Teague Road was responsible for both reversible and irreversible health damage. To gain a better perspective on the plausibility of this cause and effect relationship, we compared the findings of these health surveys with the predicted adverse health impacts suggested from exposure estimates and toxicological studies on laboratory animals.

Table 3. Toxicological Effects in Animals for Those Chemicals Identified in Contaminated Well Water Along Toone-Teague Road.

	Cancer	Reproductive Effects	Kidney Effects	Liver Effects	Central Nervous System Effects	Other
carbon tetra-chloride (18,700)[a]	+	+	+	+	+	Optic Nerve
chloroform (1890)	+	+	+	+	+	
tetrachloro-ethylene (2400)	+	+	+	+	+	
benzene (15)	+	−	−	−	−	
chlorobenzene (40)	−	+	+	+	+	Blood
hexachloro-butadiene (2)	+	?	+	−	+	
hexachloro-ethane (5)	+	+	+	+	+	
hexachloro-norbornadiene	?	?	?	?	?	
naphthalene (?)	−	−	+	+	−	Eye
toluene (52)	−	−	?	?	+	
hexachloro-cyclopentadiene	?	?	+	+	+	

[a]Numbers in parentheses are the highest concentrations (μg/L) detected in well water in 1978 in any well along Toone-Teague Road.

A cursory review of the toxicological literature indicates a wide range of toxic effects associated with those chemicals identified in contaminated well water along Toone-Teague Road (Table 3). Of particular note are the chlorinated hydrocarbons, carbon tetrachloride, chloroform, tetrachloroethylene, and hexachloroethane, all of which have demonstrated carcinogenic effects, reproductive impairment, and damage to the liver, kidney, and central nervous system in animal bioassays. Because carbon tetrachloride was present in most well water at concentrations exceeding tenfold the concentration of the next most abundant contaminant (chloroform), we have focused on the risk associated with this chemical, although exposure estimates have been made for several other contaminants.

Carbon tetrachloride is a volatile chlorinated solvent commonly used as a dry cleaning and degreasing agent. In Velsicol's Memphis plant, it was used as a solvent in the manufacture of chlorinated pesticides. It is a colorless liquid with a molecular weight of 154, specific gravity of 1.6 at 25 C, and vapor pressure of 89.5 mm Hg at 20 C.[13] Its solubility in water is low (0.8 g/liter at 25 C), but it is soluble in most organic solvents.

Carbon tetrachloride is readily absorbed after inhalation or ingestion and is excreted primarily through the lungs. In humans exposed to carbon tetrachloride, the most common effects are liver and kidney damage.[14] Carbon tetrachloride also acts as a central nervous system depressant.[13] Signs and symptoms of carbon tetrachloride toxicity in humans include dyspnea; cyanosis; proteinuria; hematuria; jaundice; hepatomegaly; optic neuritis and atrophy; ventricular fibrillation; eye, nose, and throat irritation; headache; dizziness; nausea; vomiting; abdominal cramps; and diarrhea.[14]

Enlarged or fatty liver and elevated SGOT, serum bilirubin and other indicators of liver dysfunction are observed in most cases of symptomatic carbon tetrachloride poisoning in humans.[13] Such effects may occur with chronic exposure to carbon tetrachloride at 5 to 115 ppm (32 to 735 mg/m^3) in the air.[13] Changes in biochemical indicators of liver function such as SGOT and serum bilirubin occur even at exposures where no symptoms are reported.[13] Reported values of oral LD_{50} in rats have ranged from 2.8 g/kg[15] to 6.4 mL/kg (about 10 g/kg).[14] Because of the ease with which carbon tetrachloride causes fatty liver and liver necrosis in animals, it has been widely used to study effects in different species, and to follow the biochemical and cellular events leading to liver damage.[16-17]

Several studies have examined the effects of subchronic and chronic exposure of animals to carbon tetrachloride. Prendergast et al.[18] exposed groups of guinea pigs, rats, rabbits, dogs, and squirrel monkeys to carbon tetrachloride at 515 mg/m^3 (82 ppm) 8 hours/day, 5 days/week over 6 weeks, and at 61 (10 ppm) and 6.1 mg/m^3 (1 ppm) continuously for 90 days. At the two higher doses, liver damage was noted in all species. Guinea pigs and rats showed fatty changes, fibroblastic proliferation, collagen deposition, hepatic cell degeneration and regeneration, and structural alteration in the liver lobules at the two highest dose levels. No adverse effects were seen at the lowest dose level. Adams et al.[19] observed minimal signs of liver damage in guinea pigs exposed to carbon tetrachloride at 5 ppm, 7 hours/day, 5 days/week, for 258 days. This is an average daily dose of 1.5 mg/kg/day, considering that lung absorption in guinea pigs is about 60% of inhaled material.

In humans, the vapor is slightly irritating to the eyes, and systemic absorption may cause serious and even fatal systemic poisoning. Of particular relevance to Rhamy's health survey is the repeated observation that carbon tetrachloride may affect vision by causing optic neuritis and optic atrophy, and that constriction of the visual fields is an early sign of poisoning.[12,15,20]

Several studies in mice, rats, and hamsters have demonstrated that carbon tetrachloride is carcinogenic in those species and, hence, may present a carcinogenic risk to humans. Reuber and Glover[21] treated groups of five strains of rats with carbon tetrachloride at 1.3 mL/kg body weight twice per week by subcutaneous injections for life. Those strains most sensitive to the toxic effects of carbon tetrachloride (Sprague-Dawley and Black) died after an average of 11 to 13 weeks with severe liver cirrhosis. Those strains that survived longest (Japanese and Osborne-Mendel) showed less severe liver cirrhosis, but developed a high incidence of liver carcinoma (up to 80%). Edwards[22] found hepatomas in all of 54 strain A mice and 126 of 143 strain C3H mice given 0.1 mL of a 40% solution of carbon tetrachloride in olive oil by gavage "two or three" times weekly for 23 to 58 treatments. The mice were necropsied 2 to 147 days after the last treatment. One of 23 C3H control mice and none of 22 control strain A mice developed hepatomas. Edwards et al.[23] treated strain L mice with 0.1 mL of a 40% solution of carbon tetrachloride in olive oil "usually three but occasionally two times weekly" for a total of 46 treatments over 4 months. Of 73 treated mice, 34 developed hepatomas compared to 2 of 152 control mice.

In a study by Eschenbrenner and Miller[24] which examined the effect of differences in dose level and interval between doses on liver hepatoma induction in Strain A mice, the maximum incidence of tumors (12/15) occurred in animals given 0.005 mL per gram body weight of a 2% solution of carbon tetrachloride (about 160 mg/kg) every 4 days over 116 days. These authors report a 40% incidence of liver tumors in mice (none in controls) receiving a cumulative dose of carbon tetrachloride of 1200 mg/kg.

NCI utilized carbon tetrachloride as a positive control in its carcinogenicity bioassays of several halogenated hydrocarbons.[25] Carbon tetrachloride was administered by gavage at 2,500 and 1,250 mg/kg/day for mice, and 100 and 150 mg/kg/day for rats, 5 days/week, for 78 weeks. The treated mice showed an almost 100% incidence of liver tumors and a high incidence of adrenal adenomas and pheochromocytomas.[25] Only 21% of the low-dose mice and 6% of the high-dose mice survived 78 weeks and only 1 mouse (a high-dose female) survived to the end of the study at 91 to 92 weeks. The rats also showed an elevated incidence of liver tumors. Della Porta et al.[26] also found carbon tetrachloride was carcinogenic in hamsters.

To determine whether or not adverse health impacts in humans might be expected from the carbon tetrachloride exposures associated with contamination of groundwater along Toone-Teague Road, an estimate of the exposure was made. These estimates included exposure through the consumption of contaminated drinking water, exposures from breathing carbon tetrachloride outgassed from the water during showers, and absorption of carbon tetrachloride through the skin by infants bathing in the contaminated water. These assumed concentrations of carbon tetrachloride in the contaminated well water were based on the model estimates (Figure 4) for years prior to 1978, and on the measured concentrations of carbon tetrachloride in the well water during 1978,[7] the last year of use.

Exposure estimates for four chlorinated organics, including carbon tetrachloride, were calculated for a male infant (infant A) and an adult male (adult A) who lived in a house (house A) served by a well (well A) located approximately 1800 feet north of the northern boundary of the landfill.

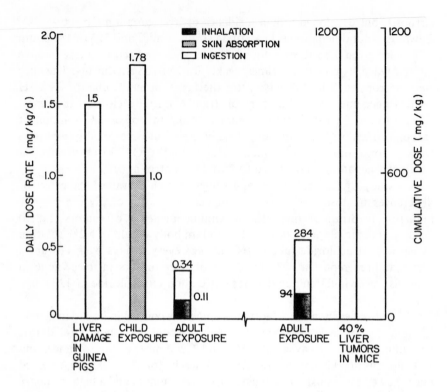

Figure 4.   Comparisons of estimated exposures to carbon tetrachloride by residents along Toone-Teague Road and exposures that product animal toxicity.

Estimating the exposure to the infant was complicated by the frequent moves the mother and child made during the period of the infant's exposure (May, 1976 to September, 1978). The exposure history was provided through Rhamy's interviews with the mother. The exposure estimates (Table 4) were based on the infant's in utero exposure and consumption of well water in formula and absorption of contaminants through the skin from the well water during baths. The infant was born on September 1, 1976, and spent much of his time away from the community after December, 1976. He visited in home B for a period during 1978, and consumed water from well B during this period. Well water was not used for drinking after March, 1978, although it was used for bathing and other domestic uses through September, 1978.

Exposure estimates for the adult male, who lived in house A, were based on consumption (2 liters/day) of the contaminated well water and inhalation of the carbon tetrachloride outgassed from the water during showers

Table 4. Estimated Chemical Doses of Selected Chlorinated Hydrocarbons to Infant A (May 1976 to March 1978)[a,b]

| | Well Water Contaminants | | | |
	Carbon Tetrachloride	Chloroform	Chlorobenzene	Tetrachloro-ethylene
Drinking Water Exposure				
Cumulative (mg/kg)	78	6.8	0.16	0.087
Average daily[c] (mg/kg/day)	0.78	.068	0.0015	0.00083
Skin Absorption[d]				
Cumulative (mg/kg)	95	8.3	0.18	0.10
Average daily[c] (mg/kg/day)	1.0	0.091	0.0021	0.00011
Total Estimated				
Exposure (mg/kg)	173	15.	0.34	0.19

[a]An average body weight of 3 kg was assumed for the period September, 1976 to December, 1977; 7 kg from March, 1978, to September, 1978.

[b]Calculations are based on in utero exposure from mid-May, 1976, through birth (assumed mother consumed 1 liter/day of well A water during a period her average body weight was 60 kg); water consumption by child of 0.5 liter/day and three baths per week from birth (September, 1976) through November, 1976, at house A. Child did not live in the community after December, 1976, and until March, 1978. Assumed water consumption by child of 1 liter/day for a total period of four days from well B in March, 1978; assumed 14 baths during period from March to September, 1978, (drinking prohibited in March, 1978).

[c]Average daily exposure for highest year (1977 for ingestion, 1978 for skin absorption) in mg/kg/day.

[d]Assumed child bathed in 30 liters/day of well water, coming into contact with 50% of the water from which 10% of the contaminant was absorbed through the skin.[27]

(Table 5). The concentration of carbon tetrachloride in bathrooms during showers was based on the assumption that the carbon tetrachloride in the bathroom air would quickly reach equilibrium with the carbon tetrachloride in the water during a shower. No estimates were made of exposures associated with other uses of the water, such as cooking and washing the dishes.

The estimated exposures to carbon tetrachloride for both the infant (Table 4) and the adult (Table 5) are contrasted in Figure 4 with the minimum daily exposures producing liver damage in guinea pigs and the lifetime cumulative exposure producing liver cancer in mice. As can be seen, the estimated daily exposure in 1977 to infant A exceeded the minimum effective dose necessary to cause liver damage in guinea pigs. A similar exposure estimate for adult A is within a factor of five of the minimum effective dose for liver damage in guinea pigs.

It should be pointed out, however, that these daily exposure rates were based on the predicted year *average* exposure during the highest year of exposure (1977 for house A). However, monitoring data collected from 1978 on indicate that the concentration of carbon tetrachloride varies over an approximate order of magnitude around the mean (e.g. at house A in 1978, the carbon tetrachloride concentration ranged from approximately

Table 5.   Estimated Chemical Doses of Selected Chlorinated Hydrocarbons to Adult A (1969 to 1978)[a,b]

	Well Water Contaminants			
	Carbon Tetrachloride	Chloroform	Chlorobenzene	Tetrachloro-ethylene
Drinking Water Exposure[c]				
Cumulative (mg/kg)	190	16	0.37	0.22
Average daily[d] (mg/kg/day)	0.23	0.022	0.00040	0.00023
Inhalation Exposure[e–g]				
Cumulative (mg/kg)	94	—	—	—
Average daily[d] (mg/kg/day)	0.11	—	—	—
Total Estimated				
Exposure (mg/kg)	284	16	0.37	2.2

[a]Body weight assumed to be 60 kg.
[b]Calculations are based on drinking water consumption through November, 1977 and inhalation exposure through March, 1978.
[c]Assumed 2 liters/day of well water consumed.
[d]Average exposure for highest year in mg/kg/day.
[e]In order to estimate carbon tetrachloride concentrations in the bathroom during and immediately following showering, the following assumptions were made:

(1)   Water temperature was 50 C.
(2)   Henry's law constants are calculated from:

$$H = \frac{P^*_{A^1}}{P^*_{A^2}} = 16.04 \frac{P^\circ_A M_A}{T P^*_{A^2}}$$

where $P^*_{A^1}$ and $P^*_{A^2}$ are the equilibrium concentrations of CCl_4 in the air and water phases, respectively

P°_A is the vapor pressure of pure CCl_4 in mm Hg
M_A is the molecular weight of the CCl_4
T is the temperature in $^\circ K$
P_{A^2} is the solubility of the CCl_4 in water in mg/L

Therefore:

$$H = \frac{P^*_{A^1}}{P^*_{A^2}} = 16.04 \frac{P^\circ_A M_A}{T P^*_{A^2}} = \frac{(16.04)(314)(153.9)}{(323)(875)} = 2.7$$

(3)   The vapor pressure of carbon tetrachloride at 50°C was calculated to be 314 mm Hg using the Antoine equation.[28]
(4)   Solubility at 50 C was extrapolated from solubilities at 20 C, 25 C, and 30 C.
(5)   Total water use per shower of 100 liters.[29]
(6)   Bathroom dimensions of 6' × 6' × 8' = 8,200 liters.
(7)   Mixing assumed complete and instantaneous within this air volume.

[f]Assumed one hour per week was spent in the contaminated air and the volume breathed was 28.6 liters/minute (17.1 breaths/minute).[30] According to EPA,[31] 40% of the inhaled material (carbon tetrachloride) is absorbed.
[g]Using assumptions similar to those above, exposure via inhalation to chloroform, chlorobenzene, and tetrachloroethylene are found to be insignificant, relative to carbon tetrachloride.

2,000 ppb to 18,000 ppb, with a mean of 9,000 ppb). Therefore, the maximum daily exposure rate over some portion of 1977 was probably considerably higher than those estimates presented in Tables 4 and 5.

In comparison to the cumulative exposures producing 40% liver tumors in mice, the estimated cumulative exposure to adult A from approximately 1969 through 1977 (see Figure 3) suggests that adult A has been exposed to approximately 20% of the carbon tetrachloride that was demonstrated to produce a 40% incidence of liver tumors in mice.

Based on the comparisons in Figure 4, if infant A were as sensitive to the hepatic effects of carbon tetrachloride as the guinea pig, then infant A should have experienced liver damage. In fact, Rhamy's physical examination of this child at age six (approximately four years after exposed supposedly ceased) revealed an enlarged liver, and elevated blood levels of alkaline phosphatase, SGPT, SGOT, and LDH, all suggestive of liver damage. Furthermore, the child suffered from upper respiratory irritation, nausea, vomiting, black tarry stools, frequent headaches, dizziness, bloodshot eyes, and photophobia during the time of exposure to carbon tetrachloride and the other contaminants in the well water (age 0 to 2 years).

During the time of maximum exposure (1975 to 1978) adult A was hospitalized with a kidney infection, experienced lethargy, shortness of breath, frequent headaches, ringing in the ears, decreased vision, nausea, vomiting, and soreness of the throat. On physical examination in 1982, Rhamy observed evidence of kidney dysfunction and an enlarged liver.

CONCLUSIONS

Chemical wastes from Velsicol Chemical Corporation's Hardeman County landfill have seeped into the water table aquifer and contaminated private wells along Toone-Teague Road. Based on a hydrogeological model and groundwater monitoring since 1978, local residents appear to have been exposed to progressively increasing concentrations of chlorinated solvents (Table 6), particularly carbon tetrachloride, from about 1969 to 1978, when an alternative water supply was provided.

Two independent health surveys conducted in 1978/1979 and in 1982 strongly suggested that the contaminated groundwater was responsible for a wide range of symptoms (e.g., headaches, nausea, upper respiratory infection, etc.) and health damage, including optic atrophy, liver enlargement, and peripheral neurological changes.

In a comparison of the estimated exposures to carbon tetrachloride by an adult and an infant residing in close proximity to the landfill, the estimated average daily exposure to carbon tetrachloride for the infant exceeded the minimum effective dose that produced liver damage in guinea pigs, and the estimated adult exposure was approximately 20% of this minimum effective dose. These estimates would suggest, considering the presence of a number

Table 6.   Indoor Air Concentrations of Selected Organic Compounds in Houses Along Toone-Teague Road, 1978[4]

Compound	Concentration Range (μg/m³)
hexachlorocyclopentadiene (3/5)[a,b]	0.006 – 0.10
heptachlorobicycloheptene (1/5)[a]	0.6
carbon tetrachloride (5/5)[b]	3.6 – 41
before shower (1/1)	23
after shower (1/1)	3600
tetrachloroethylene (4/5)	2.4 – 5.7

[a]Numbers in parentheses are numbers of houses from those examined with contaminant above detection level.
[b]ACGIH recommended occupation standards: carbon tetrachloride, 65,000 μg/m³; hexachlorocyclopentadiene, 100 μg/m³.

of other liver toxins (e.g., chloroform and tetrachloroethylene) that were present in the water but which were not included in this risk estimate, that Toone-Teague residents were at high risk of liver damage, and perhaps other adverse effects. For example, the cumulative exposure to carbon tetrachloride from 1969 to 1978, in comparison to the lifetime cumulative exposure to mice that produced liver tumors, suggests that residents are at high risk of cancer from their exposure to carbon tetrachloride.

The health surveys, exposure estimates, and risk assessment presented above are not without methodological flaws. However, we believe that the evidence is strongly suggestive that residents along Toone-Teague Road are at high risk of liver dysfunction and cancer. Although Rhamy's observation that there is an elevated incidence of optic atrophy and peripheral neurological changes in this population is plausibly related to their exposure to carbon tetrachloride, data were not sufficient to permit a quantitative risk estimate.

REFERENCES

1. Epstein, S.S., Brown, L.O., Pope, C. Hazardous Waste in America. Sierra Club Books, San Francisco, 593 pp, 1982.
2. Sprinkle, C.L. Leachate Migration from a Pesticide Waste Disposal Site in Hardeman County, Tennessee. U.S. Geological Survey, Water Resources Investigations 78–128, 1978.
3. Rima, D.R., Brown, E., Goerlitz, D.F., and Law, L.M. Potential Contamination of the Hydrologic Environment from the Pesticide Waste Dump in Hardeman County, Tennessee. U.S. Geological Survey, Administrative Report to the Federal Water Pollution Control Administration, 1967.
4. Clark, C.S., Meyer, C.R., Gartside, P.S., Majeti, V.A., Specker, B., Balistreri, W.F., and Elia, V.J. An Environmental Health Survey of Drinking Water Contamination by Leachate from a Pesticide Waste

Dump in Hardeman County, Tennessee. *Arch. Environ. Health* 37:9–18, 1982.

5. Geraghty and Miller, Inc. Groundwater Conditions in the Vicinity of a Chemical Waste Disposal Site in Hardeman County, Tennessee. Final Report prepared for Velsicol Chemical Corporation, (February 2, 1979).

6. Geraghty and Miller, Inc. Report on Groundwater Conditions, Hardeman County Landfill Site, Appendix B, prepared for Velsicol Chemical Corporation (October 1979).

7. Orr, J.R., Ziegler, F.G., and Hines, J.M. Hardeman County Landfill Site Annual Monitoring Report, ERM-Southeast, Inc., Brentwood, TN, 1982.

8. S.S. Papadopulos & Associates, Inc. Testimony by Stavros Papadopulos and Steven Larson in the United States District Court for the Western District of Tennessee, Eastern Division, Woodrow Sterling et al., vs. Velsicol Chemical Corporation. Transcript of Evidence, Volume XII, July 27, 1982.

9. Ziegler, F.G., J.R. Orr and John Hines, AWARE, Inc. Hardeman County Landfill Site Annual Monitoring Report 1981, prepared for Velsicol Chemical Corporation (November 1981).

10. Konikow, L.F., and Bredehoeft, J.D. Computer Model for Two-Dimensional Solute Transport and Dispersion in Groundwater. U.S. Geological Survey, Techniques of Water-Resources Investigations, Book 7, Chapter C2, 1978.

11. Rhamy, 1982. Testimony of Dr. Robert K. Rhamy, *Sterling et al.* v. *The Velsicol Corp.*, U.S. District Court for the Western District of Tennessee, Civil Action 78–100.

12. Smith, A.R. Optic Atrophy Following Inhalation of Carbon Tetrachloride. *Arch. Ind. Hyg.* 1:348–351, 1950.

13. American Conference of Governmental Industrial Hygienists (ACGIH). "Documentation of the Threshold Limit Values." Fourth ed. Cincinnati, OH, 1980.

14. National Research Council (NRC). Drinking Water and Health. National Academy of Sciences, Washington, DC, 1977.

15. Smyth, H.F., Smyth, Jr., H.F., Carpenter, C.P. The Chronic Toxicity of Carbon Tetrachloride; Animal Exposures and Field Studies. *J. Ind. Hyg.* 18:277–297, 1936. As cited in Reference 13.

16. Cornish, H.H. Solvents and Vapors. In: Doull, J., Klassen, C.D., and Amdur, M.O., Eds. Cassaratt and Doull's Toxicology, Second ed., MacMillan Publishing Co., New York, pp. 408–496, 1980.

17. Plaa, G.L. Toxic Responses of the Liver. In: Doull, J., Klaasen, C.D., and Amdur, M.O. Eds. Cassarett and Doull's Toxicology. Second ed. MacMillan Publishing Co., New York. pp. 206–231, 1980.

18. Prendergast, J.A., Jones, R.A., Jenkins, L.J., and Seigel, J. Effects on Experimental Animals of Long-term Inhalation of Trichloroethylene, Carbon Tetrachloride, 1,1,1-Trichloroethane, Dichlorodifluoromethane, and 1,1-Dichloroethylene. *Toxicol. Appl. Pharmacol.* 10:270–289, 1967.

19. Adams, E.M., Spencer, H.C., Rowe, V.K., McCollister, D.D., and Irish, D.D. Vapor Toxicity of Carbon Tetrachloride Determined by Experiments on Laboratory Animals. A.M.A. *Arch. Ind. Hyg. Occup. Med.* 6:50–66, 1952.

20. Wirtschafter, Z.T. Toxic Amblyopia and Accompanying Physiological Disturbances in Carbon Tetrachloride Intoxication. *Amer. J. Public Health* 23:1035–1038, 1933.

21. Reuber, M.D. and Glover, E.L. Cirrhosis and Carcinoma of the Liver in Male Rats Given Subcutaneous Carbon Tetrachloride. *JNCI* 44:419–427, 1970.

22. Edwards, J.E. Hepatomas in Mice Induced with Carbon Tetrachloride. *JNCI* 2:197–199, 1941.

23. Edwards, J.E., Heston, W.E., and Dalton, A.J. Induction of Carbon Tetrachloride Hepatoma in Strain L Mice. *JNCI* 3:297–301, 1942.

24. Eschenbrenner, A.B., and Miller, E. Liver Necrosis and the Induction of Carbon Tetrachloride Hepatomas in Strain A Mice. *JNCI* 6:325-341, 1946.

25. Weisburger, E.K. Carcinogenicity Studies on Halogenated Hydrocarbons. *Environ. Health Persp.* 21:7–16, 1977.

26. Della Porta, G., Terracini, B., and Shubik, P. Induction with Carbon Tetrachloride of Liver Cell Carcinomas in Hamsters. *JNCI* 26:855-863, 1961.

27. U.S. EPA. Office of Pesticide Program, Guidelines on Predicting Exposures to Pesticides, 1980.

28. Lange, N.A. Handbook of Chemistry, McGraw-Hill, Inc., 1967.

29. Metcalf & Eddy, Inc. Wastewater Engineering, McGraw-Hill, Inc., 1972.

30. Taylor, C. *Amer. J. Physiol.* 135:27, 1941.

31. U.S. EPA Office of Health and Environmental Assessment. Health Assessment Document for Carbon Tetrachloride. EPA-600/8-82-001, 1982.

CHAPTER 13

Problems in Determining Health Effects of a Community Exposed to Toxic Wastes

Evelyn Talbott, Lewis Kuller, Patricia Murphy, Edward Radford, and
Neal Traven

INTRODUCTION

Increasingly, citizens are becoming aware of the potential harm of hazardous waste sites near their communities. In the case of Canonsburg,
Pennsylvania, a uranium processing site now known as the Canon Industrial Park began operation in 1910. The refinery was directly adjacent to the
two communities of Strabane and Canonsburg, 18 miles southwest of Pittsburgh, with a combined population of 15,000 people. Our objective is to use
Canonsburg as an example to describe some techniques for evaluating the
health status of a community exposed to a potentially hazardous substance.
Our examples will include: (1) longitudinal mortality study, (2) cross-sectional morbidity investigation, and (3) retrospective case-control studies.

J. B. Andelman and D. W. Underhill (Editors), *Health Effects from Hazardous Waste Sites*
© 1987 Lewis Publishers, Inc., Chelsea, Michigan – Printed in U.S.A.

BACKGROUND

The refinery's founder, Joseph Flannery, was one of the first developers of vanadium steel. He had a sister who was stricken with cancer, and hearing of radium's curative powers, embarked upon the refining of uranium, and hence, radium from uranium ore. At one time, his American Standard Chemical Company produced 75% of the world's radium.

Standard Chemical and its successors gave up the radium business in Canonsburg in 1929. After a period of nonproduction, the site was purchased by Vitro Corporation to process leftover ore and extract uranium for nuclear weapons. Until the company left in 1958, radioactive waste continued to be dumped around the plant. Because depleted uranium ore was not considered hazardous in the 1950s and was not regulated by the Atomic Energy Commission, Vitro Rare Metals created a radioactive lagoon at the plant's western boundary.

EXPOSURE ESTIMATES

At the request of the Department of Energy (DOE), Oak Ridge Operations conducted radiological surveys during the period of March to July, 1977, at the Canonsburg site. Gamma levels onsite were in excess of 500 μR/hr in some places. Records show that approximately 30,000 pounds of uranium oxide were extracted from 200,000 pounds of waste received from different AEC installations. Liquid wastes were discharged into an open drain which went under Strabane Avenue (the main road) and emptied into a swamp shown in Figure 1. Shown in this figure are the three parcels of property and their maximum gamma measurements. A drainage ditch connected the swamp with Chartiers Creek which flows into the Ohio River. Solid waste accumulated in several mounds as can be seen in the darkened areas of the figure after 1911.

Offsite measurements were taken over a one mile by 2.5-mile area. Elevated gamma levels were detected within a 1/4- to 1/3-mile radius of the plant. These aerial surveys indicate background at 3 to 7 μR/hr with exposure rates of 9 to 50 μR/hr in the vicinity of the plant. In addition to this exposure, there was also the problem of contaminated articles (bricks, boards, landfill) being taken from the Vitro Plant and the Standard Chemical Plant over the 50-year period. These made their way into the homes of the residents.

Offsite readings taken in each of the two communities of Strabane and Canonsburg indicate there are over 100 homes which have reportedly localized areas of contamination. It was because of this situation that the DOE and U.S. EPA are currently conducting a very extensive offsite monitoring effort. Presently, lot by lot gamma radiation monitoring conducted by Oak Ridge for DOE is available for only a few streets. Based on these few data

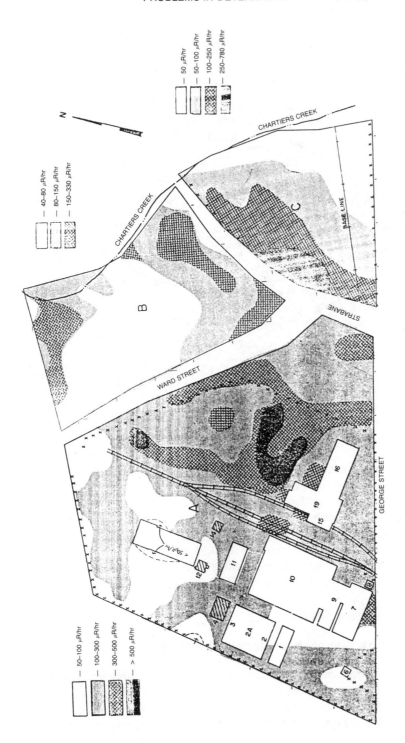

Figure 1.   Profile of average gamma radiation levels at 1 m above the surface in Parcel A, B and C.

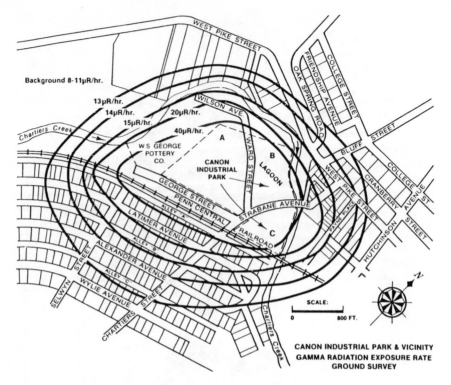

Figure 2. Canon Industrial Park and vicinity gamma radiation exposure rate ground survey.

points it appears the exposure levels range from 15 to 40 μR/hr (See Figure 2). We cannot comment on the long-term exposure of the remainder of the population until more data are made available.

Onsite indoor radon levels were quite high. Levels ranged from 10 pCi/L to 200 pCi/L. In comparison, offsite outdoor levels were low. These were typically 1.5 to 2 times background (or 0.35 to 0.75 pCi/L). Levels indoors and offsite are just now becoming available from DOE, but also appear to be in the 0.21 to 1.5 pCi/L range, which is considered background.[1]

STUDY DESIGN APPROACHES

Longitudinal Approach for Mortality Determination

As more information was released regarding the possible radiation exposures to the community at large, residents began to express their fear of excess cancer mortality. Earlier in the year a cross-sectional analysis was completed of the death rates and incidence of cancer in this area. No excess of either cancer or total mortality was reported. However, looking at the problem in such a general way could obscure a slight but significant increase within a subset of the population. Factors such as migration in and out of the area, the lag time between exposure and disease, and the very large geographic area included must all be considered when trying to determine if there is an association between disease and some environmental exposure. This is not possible with such a broad cross-sectional approach.

We believed that a longitudinal study would determine if there was a significant (10%) increase in cancer mortality. Our population consisted of 6000 individuals who had been identified through 1938 tax records as living in the Strabane/Canonsburg area at that time. The control group, chosen because of similar ethnic and socioeconomic makeup, would consist of 6000 individuals identified from South Fayette Township tax records for the same period. The present status of all individuals in each group would be determined, and death certificates for deceased individuals were to be obtained. Tracing of individuals would be conducted by utilizing long-time residents of the area to aid in identifying the cohort. Direct contact would be made to verify the individual's address and knowledge of other family relatives included in the sample. For those individuals who are not identified by usual followup methods, more extensive tracing including use of driver license records, detailed searches of death certificate files, use of property tax records, etc., could be undertaken.

SAMPLE SIZE ESTIMATES

Computation of Expected Number of Deaths

A modified Life Table Method (computer simulation) was used to generate a crude estimate of the expected number of deaths in a population cohort for the period 1940 to 1978.

The expected number of deaths was generated by applying age, race and sex specific death rates for the United States to this population. U.S. death rates for the years 1940, 1950, 1960, and 1970 were obtained from U.S. Vital Statistics Report.

Based on the age distribution of the 1940 South Fayette Census Tract and utilizing our modified Life Table Method we generated a crude estimate of the expected number of deaths in the population cohort for the period 1940 to 78. The following projections were made based on a 10%, 20%, or 30% increase in cancer mortality in our exposed population as shown in Table 1.

As our information was available by ten-year groupings only, we must make several assumptions. It is assumed that cause-specific death rates change little during a ten-year period. Furthermore, we are not adjusting for migration in or out of the area during this period. In addition, age cohorts did not escape a particular age interval until after ten years. There-

Table 1. Projected Number of Cancer Deaths Based on a 10%, 20%, and 30% Increase in Expected Mortality[a]

	Population at Risk	Deaths Expected Total	Expected Cancer Deaths[b]	10% >	20% >	30% >
			All Neoplasms			
Exposed Population	6000	3331	438	482	526	569
			Radiosensitive			
Exposed Population	6000	3331	294	323	352	381

[a]Neoplasms only. South Fayette Township—1940 age distribution U. S. Age specific mortality rates: 1940, 1950, 1960, 1970.
[b]Lung, gastrointestinal, breast, urinary tract, and hematopoietic comprise 67% of the cancers.

Table 2. Estimates of Population Size Based on 10% and 15% Increases in Cancer Mortality for Strabane Population (1940 to 1970 Followup)

α	Projected 10% Increase $1 - \beta$	n
	.60	9,527
.05	.70	12,427
	.80	16,230
	.60	6,259
.10	.70	8,646
	.80	11,801
α	Projected 15% Increase $1 - \beta$	n
	.60	4,364
.05	.70	5,693
	.80	7,436
	.60	2,867
.10	.70	3,961
	.80	5,434

fore, all age cohorts shift to higher age intervals three times in this analysis: 1950, 1960, and 1970. Thus, a person who is 15 to 24 in 1950 does not become 25 to 34 until 1960. The resulting tables probably reflect an underestimate of the expected number of deaths. Also, many cancer rates have risen markedly in the past ten years. This is not reflected in 1970 rates.

Table 2 describes the estimated population size that would be necessary for a 10% or 15% increase in cancer mortality to be statistically significant based on various estimated alpha and beta levels. We should be able to detect an increase in cancer mortality between 10% and 15% if it exists. It can be seen, however, that even with 6000 people followed 40 years, the power $(1-\beta)$ to detect the effects of low-dose radiation is small.

Several occupational studies including Hanford[2] nuclear workers and the Portsmouth Naval Shipyard workers[3-4] have reported much higher risks of cancer than previously had been expected. However, these studies of low-level radiation effects have been seriously criticized by significant segments of the professional community. It would appear that for this small population, to detect an increase of 10%, a longer followup period would be necessary (perhaps 60 or more years).

CROSS-SECTIONAL MORBIDITY SURVEY

A cross-sectional or prevalence survey design is particularly well-suited to this population. It involves a nondirectional or backward design of a study population. As mentioned, this type of design involves disease prevalence, not incidence, and either random sampling of the dynamic target population or inclusion of all eligible individuals.[5] After selection, all participating subjects are examined or questioned about disease status, exposure to the study factor level, and other relevant variables.

There are several important aspects of this design which should be considered prior to its use: (1) The disease under study should be relatively frequent or have a long duration (high survival rate). (2) Moreover, prevalence data cannot be used in the same way as incidence data to ascertain the direction of the relationship between the study factor and disease, if an association is found. The investigator cannot determine from the design alone whether the hypothesized cause was an antecedent or a consequence of the disease. Exposure to elevated levels of gamma rays in this case can be controlled for to some extent by imposing a length of residence ($\geq$ 15 years) requirement on those who came in for screening from the exposed communities. Similarly, to be eligible for screening in the unexposed or comparison community, the individual had to have never lived or worked within the Strabane/Canonsburg study area and must have resided in the control community ten or more years. (3) Lastly, sometimes cross-sectional studies do not use random sampling, and a serious source of bias can be introduced in not knowing the probability that an individual in the target population will be selected. Self-selection can lead to a distortion of results if either disease status or study factor levels inadvertently influence the probability of selection. Another form of selection bias which may influence results is unequal diagnostic surveillance practices between groups. Most cross-sectional studies involve sampling from only one population. In the case of Canonsburg, two separate towns had to be used. These aspects will be addressed individually in the discussion section.

Table 3.   Demographic Features of Eligible Populations

	Canonsburg/Strabane N = 514 Eligibles	Muse N = 424 Eligibles
Sex		
Males	198 (39%)	195 (46%)
Females	316 (61%)	229 (54%)
Race		
White	492 (96%)	422 (100%)
Black	22 (4%)	2 (0%)
Age		
Mean	55.52 yrs	50.63 yrs
Median	57.55 yrs	50.82 yrs
S. D.$_{\bar{x}}$	16.56	16.99
Residence		
Mean	37.47 yrs	31.22 yrs
Median	33.50 yrs	27.57 yrs
S. D.$_{\bar{x}}$	16.71	15.09
Screening Status		
Screened	310 (60%)	231 (54%)
Not Screened	204 (40%)	193 (46%)

DEFINING THE POPULATION AT RISK

In order to determine the magnitude of the population at risk, a household roster was administered which gathered information on age, race, sex, number of individuals who lived at home, and number of years lived at that particular address. In addition, information was obtained on individuals who may have lived within the confined area but had lived at several addresses. To meet the criteria for this study, the individual had to live within a 1/3-mile radius of the park and had to be a current fulltime resident, 21 years of age or older, and must have lived within the area for 15 or more years. Of 423 households in the Strabane and Canonsburg 1/3-mile area, this information was obtained for 388 or 92%. It was determined that 514 of the 850 people in the Strabane/Canonsburg study area were eligible for inclusion in the study.

The town of Muse, Pennsylvania was chosen as the control population. It is located 5 miles from the site and exhibited background levels of radiation (8 μR/hr). The socioeconomic, ethnic, and age make-up was also similar (see Table 3).

Now that the population at risk had been enumerated, the most appropriate morbidity endpoint would have to be determined. Several large epidemiologic studies have included substantial information on persons with whole body exposures under 50 rads. Recently Dreyer and Friedlander[6] reviewed this literature in order to identify health effects associated with lifetime exposure to a cumulative mean organ dose up to 1000 rads of low LET radiation. Included in this review are only those studies which identify the effects that were observed in at least two independent studies.

The lowest cumulative organ dose at which cancer has been reported is about 6 rads for thyroid cancer.[7] This effect was observed among children treated for tinea capitis by X-ray epilation of the skin. At doses of 30 rads and above, reports of acute and chronic myelogenous leukemia begin to appear. Salivary gland tumors have been observed to occur at doses of 39 rads to the parotid gland. At doses as low as 60 to 150 rads, the following health effects have been reported: intestinal cancer, brain tumors, bone cancer, and breast cancer in females.

Of the principal types of thyroid cancer, the most common are papillary and follicular accounting for 80% of all thyroid cancer. The five-year survival rate overall for men is 80% and 90% for females. Therefore, given the radiosensitive nature of the thyroid gland, the high survival rate, and the ease with which a head and neck exam could be carried out, the thyroid was chosen as the target organ. The age-adjusted rate of thyroid cancer in the United States is 3.7 per 100,000. Consequently, it would be possible to identify a small number of radiation-induced thyroid cancers as the number of background cancers expected is so small.

In addition to the significant correlation between radiation and thyroid cancer, there have also been several reports in the literature of an increased rate of nodular thyroid disease and adenomas.[8-10] A 27% incidence of thyroid nodularity was found in a group of 1000 subjects[10] irradiated primarily to the tonsils and nasal pharynx. The mean age was 33 years. Unfortunately, the exam was not done blind, and there was no control group. Other studies have reported no effect. Royce[11] reported a thyroid nodularity rate of 15.7% in exposed and 12.7% in nonexposed controls with a thyroid cancer rate of 6.1% in the cases and 4.5% in the controls. This was a blinded study involving 214 persons with verified head and neck irradiation and 243 concurrently examined controls. However, there was a selective recall bias potential as the total group of irradiated individuals actually consisted of 738 patients.

A study by Shore, Albert and Pasternak[7] has provided some data in the low-dose range in a population treated for tinea capitis. The dose to the thyroid gland was estimated to be approximately 6 rads. Followup information has been obtained on approximately 80% of the original cohort of 2,545 irradiated and 1,809 nonirradiated controls with the same disease. The rates for thyroid adenoma were 2.7/1,000 in the exposed group and zero in the controls. Among the irradiated cohort, the crude rates by sex were 7.0/1,000 in females and 2.1/1,000 for males. No carcinomas were detected.

Modan, et al.[12] have also reported on a population of about 11,000 children receiving irradiation to the scalp for treatment of tinea capitis.

Again, the dose to the thyroid was quite low (9 rads). The exposed group was paired to either a nonirradiated control matched on sex, age, country of origin, and immigration period, or a sibling matched within five years of age or both. Cases were ascertained by tracing through the Central Israeli Tumor Registry and the death certificates file. No data were available on the incidence of benign tumors.

A rate of 1.1/1,000 was found in the irradiated group for malignant neoplasms of the thyroid compared to a rate of 0.2/1,000 in both control groups. This finding was quite unexpected by the authors in view of the low dose to the thyroid gland, but several explanations are offered: (1) careless-ness during irradiation, resulting in higher exposures to the thyroid, (2) extreme sensitivity of the thyroid gland to irradiation, or (3) tumor produc-tion through a hypophyseal/thyroid axis after a high radiation dose absorbed by the hypophysis.[12]

An updated, final report on this cohort was published in 1980 by Ron and Modan.[13] Irradiation occurred at the time of immigration to Israel between 1948 and 1960, and the ages of the subjects at the time of irradiation ranged from 1 to 15 years. All subjects (exposed and unexposed) were born between 1930 and 1960 and were of Asian-African origin. Two control groups were employed: one group of 10,842 population controls and one group of 5,400 nonirradiated siblings. There was a statistically significant increase in the rate of total thyroid operations between the irradiated cohort and both control groups, but the elevated risk was shown to be limited to neoplastic conditions.

METHODS AND MATERIALS

There are very few general population surveys of thyroid disease in a free living population of essentially nonirradiated individuals. Examination of patients in a nonirradiated population for nodular thyroid disease in Tecumseh[14] yielded a 0.8% prevalence, and in Framingham the nodularity rate was 2.7%.[15] The true incidence of thyroid cancer in an irradiated population is not known because not all patients found to have palpable thyroid abnormalities have been explored. We therefore assumed a normal background prevalence rate of 3% for thyroid nodularity and adenoma and estimated the sample sizes needed to detect a three-fold excess of thyroid nodularity and thyroid cancer. It was anticipated that at least 300 individ-uals would be needed in each group. A complete (100%) sample would be necessary because of the small population in each town.

Thus, we embarked on a study to detect the prevalence of palpable thy-roid disease using standard physical diagnosis. A form letter explaining the purpose of the study and asking for the residents' cooperation was sent to each household listed in our sampling frame using cross-reference directo-ries of the involved areas.[16]

A phone call was made to each resident. A detailed list of individuals permanently residing at that address was obtained. If the criteria for the study were met and the person consented to participate, an appointment for a thyroid screening exam was made. If the criteria were not met, but the person still wished to be screened, an appointment was also made, but the information was analyzed separately. When a refusal was encountered, information was obtained on eligibility and a history of irradiation, occupation, education, and history of thyroid disease.

PROCEDURES AT THE CLINIC

To ensure the minimum amount of bias associated with the physician's a priori knowledge of the screening participant's residence, both cases and controls were scheduled whenever possible for the same screening clinic date. This blinded method helped reduce criticism in the study design that the screening physicians were more likely to find significant disease and nodularity in an exposed compared to a nonexposed group of individuals. Also, the residential histories were kept from the clinical form, and the screenees were cautioned not to divulge their addresses or length of residence. The clinics were held once a week. We were fortunate to enlist the aid of two endocrinology fellows from the University of Pittsburgh Medical Center to perform the clinical examination of the thyroid. Two interviewers were also employed.

The questionnaire consisted of requesting information on age, residential history, occupation, present and past medical disorders, illness, previous forms of X-ray therapy to the head and neck region, education, marital status, and a general family history.

Screening for past or present thyroid abnormalities was done both by history and physical examination. If there was a positive result on physical examination without a previous history of thyroid disease, the person was referred to his own physician or to the university hospital clinic for a complete thyroid workup, including a scan of thyroid activity. For those with a history of thyroid disease or previous surgery, medical records were obtained, and the accuracy of the diagnosis was confirmed by the endocrinologists on the team. A total of 286 persons were screened from Strabane/ Canonsburg and 224 from Muse.

Prior to the study, the possible radiation-related thyroid abnormalities were designated as: thyroid carcinoma, thyroid adenoma confirmed by surgical-pathologic examination, and a solitary nodule without histologic confirmation but preferably with evidence by scan of nonfunctioning tissue. All other thyroid abnormalities found were considered to be unrelated to radiation, including diffuse goiter, multinodular goiter, Hashimoto's thyroiditis, or hypothyroidism.

RESULTS

A total of 286 individuals from Strabane/Canonsburg and 224 from Muse, Pennsylvania, participated in the Canonsburg health effects study. This represents 60% of the eligible individuals, as determined by a household interview survey. The nonrespondents were also queried via telephone interview regarding age, residential history, education, if they have ever worked at Vitro, medical X-ray history, and history of thyroid disease. The age distribution of the two screened populations was similar (Table 4).

The comparability of the two groups is very important from the standpoint of the distribution of thyroid disease in general. The salient demographic factors were as follows: the ethnic origin of eligible participants was similar with 50% in Strabane/Canonsburg and 62% in Muse, having parents who were born in this country. The only notable difference concerned the country of birth of mother and father. There were slightly more individuals in Strabane/Canonsburg whose mother and father came from Yugoslavia (approximately 14% vs. 3% in Muse). A greater proportion of Muse inhabitants had ancestors from Italy (17% in Muse compared to 9% in Strabane/Canonsburg). The other salient demographic variable was residential history which was similar with 35.5 years mean residence in Strabane/Canonsburg vs. 32 years in Muse. Education was also very similar, as was marital status (Table 5).

The age distribution of all eligible men and women in Muse and also Strabane by screening status is presented in Table 4. For men in both Strabane and Muse, the age distribution was similar across screening status categories. However, among men and women there is a greater proportion of not screened in the older age groups (greater than 60) in both Muse and Strabane/Canonsburg. This difference appears uniform for both communities and suggests little effect of bias in selection for screening.

Proportion screened by proximity to the industrial site also indicated little difference in response rates. Older individuals were less likely to come in for screening regardless of proximity to the plant.

Although not all eligible people in the two communities agreed to be examined, 88% were contacted via telephone interview. The rates of thyroid disease among those who did not come in for screening are very low in men and women, 3.7% and 3.6% among exposed and nonexposed women, and none reported among men in either groups.

Males, historically, are reported to have a much lower prevalence of thyroid disorders than women. In the exposed community, a total of 5 men out of 101 who came in for screening and 8 of 75 control male participants had an abnormal exam (Table 6). This small number precluded any in-depth analysis. Also, significant numbers of men had to be excluded because of occupational involvement at the site. Therefore, bias in the male population cannot be ruled out. As most of the women in the area do not work outside

Table 4.   Distribution of Eligible Strabane/Canonsburg and Muse Individuals by Age and Screening Status

Strabane/Canonsburg

Age (years)	Males			Females		
	No. Eligible	No. Screened	Percent Screened	No. Eligible	No. Screened	Percent Screened
21–34	27	12	44.4	43	27	62.7
35–49	30	21	70.0	39	28	71.8
50–59	48	30	62.5	71	55	77.5
60–69	45	30	66.7	79	51	64.5
70+	25	8	32.0	61	24	39.3
Unknown	4	–	–	11	–	–
Total	179	101	53.4	304	185	60.8
	Mean age: 53.6 ± 13.6			54.0 ± 14.9		

Muse

Age (years)	Males			Females		
	No. Eligible	No. Screened	Percent Screened	No. Eligible	No. Screened	Percent Screened
21–34	39	19	48.7	37	31	83.8
35–49	50	23	46.0	64	41	64.1
50–59	42	10	23.8	51	40	78.4
60–69	26	15	57.7	34	19	55.9
70+	28	12	42.8	35	18	51.4
Unknown	2	–	–	3	–	–
Total	187	79	42.2	241	149	61.8
	Mean age: 49.4 ± 17.7			48.8 ± 15.6		

Table 5. Salient Demographic Variables of Screened Eligible Participants by Residence

Length of Residence	Strabane/Canonsburg N = 286		Muse N = 228	
	Number	Percent	Number	Percent
< 25 years	101	35.3	92	40.3
26–39 years	86	30.1	62	27.2
40+ years	99	34.6	74	32.5
mean years	35.5 years		31.6 years	
Education				
8th grade	96	33.8	62	27.2
9th to 12th	156	54.9	134	58.8
12+	32	11.3	32	14.0
Marital Status				
married	187	65.4	161	70.6
single	51	17.8	25	11.0
widow	39	13.6	31	13.6
other	9	3.1	11	4.8

the home, the problem of additional occupational exposure to females did not arise. The results for the women are shown in Table 7. The overall prevalence of neoplastic thyroid disease (single nodular, adenoma, and carcinoma in the exposed vs. nonexposed residents) was 14 out of 185 vs. 5 out of 149. A chi-square test with one degree of freedom was carried out. This was marginally significant at $p = .049$ (one-tailed). This is being reported as marginal because among the 14 cases, there is one unconfirmed history of irradiation in childhood. A relative risk of 2.3 with 90% confidence intervals of 0.95–4.9 was observed. This represents the rate in exposed divided by the rate in the unexposed (7.6/3.3).

Power estimates were recalculated based on the actual rates of nodularity found in our two populations. This is important because our previous estimates were based on only hypothesized rates. Our power was 70% to detect a twofold difference between the two populations.

An important question which arises is the possible difference in the proportion of cases in Muse and Strabane who were newly diagnosed at screening versus the previously diagnosed cases. It is possible that self-selection of the sample in Strabane/Canonsburg resulted in an increased number of preexisting cases participating in screening. Table 8 presents the final diag-

Table 6.   Total Thyroid Disease in Males

	Strabane/Canonsburg	Muse
Number of men screened	101	75
Neoplastic Thyroid Conditions		
solitary nodule	1[a]	3[b]
adenoma	1	–
thyroid cancer	–	–
Total	2	3
Other		
diffuse goiter	1	3
multinodular goiter	–	–
thyroiditis	1	–
history of hypothyroidism	1	–
thyroidectomy (Grave's)	–	2
Total	5	8
18-Month Period Prevalence Rate	[c]4.9%	[c]10.7%

[a]Worked hauling materials for Vitro for 2 weeks.
[b]2 of 3 have sought no further followup.
[c](p > 0.05; not significant.)

nosis prior to screening status. Twenty-three of forty-four documented abnormal cases (52.3%) were diagnosed prior to screening in Strabane compared to 14 of 39 (36%) in Muse. This represents an overall rate of preexisting abnormality of 12% vs. 9% respectively. This is two to three times higher than the low rate of previous disease reported among telephone interviewed nonrespondents (approximately 4% in both towns). In the neoplasia category, there is little difference with regard to the number of newly diagnosed cases — 87.5% of nodules in Strabane (7) and 100% of those in Muse (4) — were not previously diagnosed. In addition, all of the adenomas and carcinomas in both communities were previously medically treated and diagnosed.

Another criticism of this screening program may be that the control or unexposed individual was not followed up as aggressively or that appropriate tests were not conducted during his followup. Table 9 indicates the proportion of confirmed abnormal thyroid findings among exposed and nonexposed populations by diagnosis. The proportion of nonvalidated cases is comparable (17.0% vs. 22.5%).

The results were analyzed by those with length of residence 35 years or more vs. less than 35 years (Table 10). For those women living less than 35 years in each community, the rates of possible radiation-related thyroid

Table 7.   Total Thyroid Disease in Women

	Strabane/ Canonsburg	18-month Period Prevalence	Muse	18-month Period Prevalence
Number of Women Screened	185		149	
Thyroid Neoplasia		rate/100		rate/100
solitary nodule	8		4	
adenoma confirmed	5		1	
thyroid carcinoma	1		0	
Total	14	7.6[a]	5	3.3[a]
All Others				
diffuse goiter	8		13	
multinodular goiter	4		8	
thyroiditis	6		4	
history of hypothyroidism	10		3	
history of thyroidectomy for toxic goiter	5		7	
Total	33	17.8	35	23.5
All thyroid abnormalities	47	25.4	40	26.8

[a] x^2 = 2.73, one-tailed p = 0.049   RR = 2.3 (.95–4.9); (90% C.I.)

disease were 4.5% for Strabane/Canonsburg and 3.7% for Muse. For those resident more than 35 years, their rates were 11.7% for exposed vs. 3.0% for unexposed. This is significant at p = 0.055. Thus, the effect was almost entirely observed for long-term residents only, as would be expected on a radiogenic basis. This may be related to early age at exposure. Further age-adjustment procedures will also be carried out. The distribution of abnormality in the remaining three strata is unremarkable.

CONCLUSION

The rate of possible radiation-related thyroid disease in Strabane/ Canonsburg was 2.3 times higher than our control community. Because of the possibility of selection bias and the nonblinded nature of the followup, we can only state that this study is marginally significant. A followup study is planned which will repeat the reading of scans and pathology. This is, however, the magnitude of effect we might expect given the low-level nature

Table 8. Final Diagnosis in Strabane vs. Muse by "Diagnosis Prior to Screening Status" (Females Only)

Final Diagnosis	Strabane/Canonsburg N = 44						Muse N = 39					
	Not Previously Diagnosed		Previously Med. Documented		Total N 144		Not Previously Diagnosed		Previously Med. Documented		Total N 39	
	no	%	no	%	N		no	%	no	%	N	
I. Possible radiation-related												
single nodule	7	87.5	1	12.5	8		4	100	—	—	4	
adenoma	—	—	5	100	5		—	—	1	100	1	
thyroid cancer	—	—	1	100	1		—	—	—	—	—	
II. Goiter (simple & simple diffuse)	8	100.0	—	—	8		13	100	—	—	13	
III. Multinodular goiter	3	75.0	1	25.0	4		4	50	4	50	8	
IV. Thyroiditis	3	50.0	3	50.0	6		4	100	—	—	4	
V. Thyroidectomy	—	—	5	100.0	5		—	—	8	100	8	
VI. Hypothyroidism (documented) prior to screening[a]	—	—	7	100.0	7		—	—	1	100	1	
Total	21	47.7	23	52.3	44[a]		25	64.1	14	36%	39[a]	

[a]Three of exposed and one of nonexposed cases reported of hypothyroidism which is undocumented.

Table 9. Proportion of Confirmed Abnormal Thyroid Findings Among Exposed and Nonexposed Populations by Diagnosis

Diagnosis Properly Validated[a] Condition	N = 47 Strabane/Canonsburg			N = 40 Muse		
	#	%	Total n	#	%	Total n
1. Uninodular Goiter[b]	7	87.8	(8)	3	75.0	(4)
2. Multinodular Goiter[c]	3	75.0	(4)	5	62.5	(8)
3. Diffuse Simple & Undiff Goiter[d]	6	75.0	(8)	10 +	76.9	(13)
4. History of Thyroidectomy[e]	10	90.9	(11)	8	100.0	(8)
5. Thyroiditis[f]	5	83.3	(6)	4	100.0	(4)
6. History of Hypo-thyroidism[g]	8	80.0	(10)	1	33.3	(3)
Total Abnormals Validated	39	82.9		31	77.5	
Total Abnormals Not Validated (Confirmed)	8	17.0	(47)	9	22.5	(40)

[a] Based on appropriate followup exams/bloodwork, etc. Includes one cancer of thyroid + 4 of 10 had blood tests only.

Criteria:

[b] Thyroid scan and uptake.
[c] Combination clinical exam and scan.
[d] Thyroid function/scan and exam.
[e] Pathology/surgery reports only.
[f] Thyroid antibody levels TFTs.
[g] Thyroid function tests (historical).

Table 10. Distribution of Possible Radiation-Related Thyroid Disease in Exposed and Nonexposed Communities By Length of Residence

Length of Residence:		Not Radiation-Related	Possible Radiation-Related	Total
<35 years	Strabane/Canonsburg	103	5 (4.6%)	108
	Muse	79	3 (3.7%)	82
	Total	182	8 (4.2%)	190
		x^2 = .09	p = n. s.	
35 + years	Strabane/Canonsburg	68	9 (11.7%)	77
	Muse	65	2 (3.0%)	67
	Total	133	11 (7.6%)	144
		x^2 = 3.7	p = .055	

of the exposure. It should be recognized that the overall rate of thyroid disease was similar in the two populations.

We cannot comment on those who did not come in for screening except to state that their self-reported rate of disease was 2 to 3 times lower in both communities than among those who came in for screening. This reflects a bias in the same direction for both communities. This investigation, however, is the first of its kind to investigate the low-level exposure of a population over a long period of time, and, if possible, should be replicated in other populations to determine if there is consistency and repeatability of its results. The high prevalence of thyroid abnormality, relative potential for good clinical examination, and further diagnostic evaluation should provide a useful method for evaluating low-level gamma radiation exposure.

CASE-CONTROL STUDY

The final example presented today is that of a retrospective case-control study. In this type of study, persons with the disease of interest are compared to persons without the disease to ascertain relevant exposures in the etiology of the disease. The numerical estimator of the effect is known as the odds ratio and is directly interpretable as the relative risk obtained from a prospective study. The methodology is particularly suited to the study of rare diseases such as cancer.

In Canonsburg, the community is also exposed to excess radon gas, a decay product of radium with a half-life of about four days. Radon is ubiquitous in our environment and diffuses into the atmosphere where it undergoes a series of rapid decays into radon "daughter" products.

Radon causes lung cancer in animals, and underground miners exposed to high concentrations experience dramatic increases in rates of disease.[17] Furthermore, a small dose rate may be more effective for a given cumulative dose, meaning that background exposures may be more hazardous than those predicted from linear extrapolation of mining data.[18-19] The concern in Canonsburg is that excessive exposure to radon gas over an extended time period could lead to an increase in lung cancers in exposed residents. The average outdoor radon concentrations at the Canon Industrial Park and vicinity were determined by DOE in 1978–79.[1,20] The highest outdoor concentrations on site (0.88 to 2.4 pCi/L) exhibited a gradient corresponding to prevailing westerly wind direction. This influence of wind is also evident offsite. The excess radon concentration downwind approaches 0.5 pCi/L the inhabited area about 200 meters from the site and is additive to normal indoor concentrations. Offsite indoor measurements began in August, 1978 and are continuing.

In 1980, a case-control study was designed and conducted to assess the association between lung cancer and environmental radon exposure in the Canonsburg area.[20] Death certificates for a 30-year period were obtained for 116 lung cancer cases and 125 ASHD controls in the Strabane/Canonsburg area. As lung cancer is a rapidly fatal disease, death certificates provide a 90% detection rate of newly diagnosed incidence cases, thereby introducing almost no sampling variability. In addition, when lung cancer is

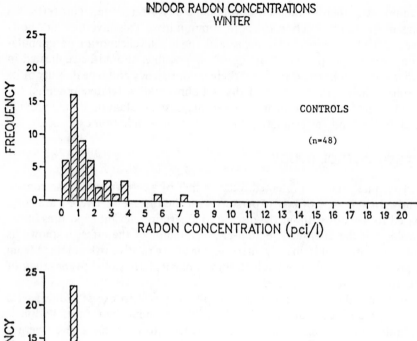

Figure 3. Indoor radon concentrations—winter.

listed on a death certificate as the underlying cause of death, it is nearly always confirmed by the hospital diagnosis, which means there would be very little chance of misclassifying lung cancer cases. Indoor radon concentrations were taken in homes of 50 lung cancer cases and 48 heart disease controls as an index of cumulative exposure. These were taken over a period of three months in the summer and three in the winter. The distribution of concentrations were similar among case and control homes as shown in Figure 3 (winter only). The range of exposures was well within background expectation as were lung cancer rates in the area. There was, furthermore, no geographic clustering of cases near the Canon site. Although an association of environmental radon exposure due to the disposal of tailings did not have a significant impact on the health of residents living in the area.

A variety of factors preclude the drawing of a firm conclusion to this study. Consent to participate was obtained for only 43% of the cases (50/116). Fear and paranoia of the residents, inability to trace the case, and the finding that the case home no longer existed all contributed to the low response rate. The small number of cases also made it impossible to control for possible confounding factors in the analysis. Repeating the study in a much larger population of lung cancer cases wherein confounders such as smoking status, occupation, and social class could be controlled might help to clarify the relationship of indoor radon and lung cancer.

SUMMARY

The purpose of this presentation was to illustrate the variety of available approaches to study the health of a community exposed to low-level radiation. The radiation-exposed town of Canonsburg, Pennsylvania, was used as an illustrative example of how several designs could be applied to a specific study population. In general, the particular study design chosen will depend on many factors. The size of the population available for study, the type of exposure present, and the type and magnitude of effect expected must all be considered.

Investigators must also be prepared to deal with an irate and confused community, who may or may not be willing to participate in scientific investigations of their problem. Because of these constraints, the study design ultimately employed is usually a compromise between what should ideally be done and what can be done at the present. Every effort should be made to carry out the most scientifically sound study with the population and resources that are available.

ACKNOWLEDGMENT

This work supported by Cooperative Agreement CR806815 from the U. S. Environmental Protection Agency.

REFERENCES

1. U.S. Department of Energy Formerly Utilized MED/AEC Sites Remedial Action Program - Survey of Former VITRO Rare Metals Plant, Canonsburg, PA, April, 1978 - Under Contract for Oak Ridge, Division of Contract Technology, Washington, DC 20545, #W-7405-ENGL 2 DOE, EV - 00513 UC 70.
2. Najarian, T. and Colton, T. Mortality from leukemia and cancer in shipyard nuclear workers. *Lancet. i*, 1018–1020, 1978.

3. Rinsky, R. A., Zumwalde, R. D., Waxweiler, R. J., et al. Cancer mortality at a naval nuclear shipyard. *Lancet. i*, 231–235, 1981.
4. Mancuso, T. F., Stewart, A. and Kneale, G. Radiation exposures of Hanford workers dying from cancer and other causes. *Health Physics* 33, 369–385, 1977.
5. Epidemiologic Research. Kleinbaum, D., Kupper, L. and Morgenstern, H. Lifetime Learning Publications, Belmont, CA, pp. 62–68, 1982.
6. Dreyer, N. A. and Friedlander. Identifying health risks from low-dose ionizing radiation. *APHA*, June, 1982, Vol. 72:6.
7. Shore, R., Albert, R. and Pasternak, B. Long term effects of x-ray therapy for ringworm of the scalp. Society of Epidemiologic Research, 13th Annual Meeting, Minneapolis, MN, June 18, 1980.
8. Hempelmann, L. H., Hall, W. J., Phillips, M., et al. Neoplasms in persons treated with x-rays in infancy: fourth survey in 20 years. *J. Natl. Cancer Inst.* 55:519–530, 1975.
9. Favus, M. J., Schneider, A. B., Stachura, M. E., et al. Thyroid cancer occurring as a late consequence of head and neck irradiation. *N. Eng. J. Med.* 294:1019–1025, 1976.
10. Refetoff, S., Harrison, J., Karanfilski, B. T., et al. Continuing occurrence of thyroid carcinoma after irradiation to the neck in infancy and childhood. *N. Eng.J. Med.* 292:171–175, 1975.
11. Royce, P. C., MacKay, B. R., and DiSabella, P. Value of postirradiation screening for thyroid nodules. *JAMA* 242 (24), December, 1979.
12. Modan, B., Mart, H., Baidatz, D., et al. Radiation-induced head and neck tumors. *Lancet* 277–279. February, 1974.
13. Ron, E., and Modan, B. Benign and malignant neoplasms after childhood irradiation for tinea capitis. *J. Natl. Cancer Inst.* 65:7–11, 1980.
14. Matovinovic, J., Hayner, N. S., Epstein, F. H. et al. Goiter and other thyroid disease in Tecumseh, MI, *JAMA* 192:234–240, 1965.
15. Vander, J. B., Gaston, E. A., Dawber, T. R. Significance of solitary non-toxic thyroid nodules: preliminary report. *N. Eng. J. Med.* 251:970–973, 1954.
16. Dickman Criss-Cross Directories, 1980, 1981 and 1982. Dickman Directories Inc. 6145 Columbus Pike, Del, OH, 43015.
17. Kunz, E., Seve, J., and Placek, V. Lung cancer mortality in uranium miners. *Health Physics* 38, 578–580, 1978.
18. Kunz, E., Placek, V., and Horacek, J. Lung cancer in man in relation to different time distributions of radiation exposure. *Health Physics* 36, 699–706, 1979.
19. Lanes, S. Lung cancer and environmental radon: a case-control study. Doctoral Dissertation, University of Pittsburgh, Pittsburgh, PA, 1982.
20. U.S. Department of Energy. Results of Aerial Radiological Survey, Canonsburg, PA. April, 1978, EG and G Energy Measurements Group Contract EY-76-C08-1183.

APPENDIX A

The Chemical Substances Information Network

The most important public databases on chemicals can be accessed via the Chemical Substances Information Network (CSIN), so a discussion of CSIN provides a good overview.[26] CSIN is not a database, but rather a system that links together other databases oriented to chemical substances. CSIN accesses data on chemical nomenclature, composition, structure, properties, toxicity, production, uses, health and environmental effects, regulations, and other aspects of materials as they move through society. CSIN is used interactively. It will prompt the user for information, eliminating the need for the user to learn protocols for hundreds of different databases. There are three databases that CSIN accesses that have toxicological information on chemicals:

1. HAZARDLINE. This provides information on over 1500 hazardous workplace substances as defined by OSHA.

2. OHMTADS (Oil and Hazardous Materials/Technical Assistance Data System). This is a database from Chemical Information Services that has information on 1250 chemicals.

3. TDB (Toxicology Data Base). This is a database from the National Library of Medicine that contains information on 4500 chemicals.

Some additional databases that are available on CSIN or soon will be available are:

1. CAS ON-LINE. This is a database from the Chemical Abstracts Services containing data on over 6,000,000 chemicals.

2. TDMS (Toxicology Data Management Systems). This was developed by the National Center for Toxicological Research and is used to collect data by the National Toxicology Program. It contains in-depth toxicology information on chemical substances.

3. CICIS (Chemicals in Commerce Information System). Developed by the EPA, it will enable EPA to process and analyze health and safety studies, inventory data, and specific chemical regulations. Nonconfidential information is available to the public through CSIN.

CSIN and the databases it accesses are available for public use. It is currently used by government agencies, public interest groups, universities, and businesses. Users must make arrangements individually with vendors of those databases in the network that the user wishes to use.

APPENDIX B

Data Elements Contained on the National Priority List—Technical Data Base

Site Identification number
Site Name
Site Location
Site NPL Rank
Category Code
Site Status
Interim Priority Site
State Priority Site
Ground Water Observed Release
Net Precipitation
Permeability of the Unsaturated Zone
Physical State[1] (of waste)
Ground Water Route Containment
Toxicity and Persistence[2] (of waste)
Hazardous Waste Quantity[3]
Ground Water Use
Distance to Nearest Well and Population Served
Surface Water Observed Release
Facility Slope and Intervening Terrain
One-Year, 24-Hour Rainfall
Surface Water Route Containment
Surface Water Use
Distance to a Sensitive Environment[4]
Population Served by Surface Water with Water Intake Within

3 Miles Downstream From Facility
Air Observed Release
Reactivity and Incompatibility (of waste)
Toxicity
Population within a Four-Mile Radius
Land Use
HRS Score on the Worksheet
Ground Water Route Score Calculated
Surface Water Route Score Calculated
Air Route Score Calculated
HRS Score Calculated
Ground Water Route Score Reported
Surface Water Route Score Reported
Air Route Score Reported
Ground Water Drinking Supply
Type of Sampling
Past Response Activities
Codes for the Chemicals
Chemical Observed in Ground Water
HRS Score Reported
Fire and Explosion Mode Score
Direct Contact Mode Score

HRS Document File Date
Name of Aquifer of Concern
Depth to Aquifer of Concern
Name of the Chemical Used to
   Score
Quantity of Waste in Drums
Quantity of Waste in Tons
Quantity of Waste in Cubic Yards
Average Slope of Site
Average Slope of Intervening
   Terrain
Site Located in Surface Water
Site Separated from Water Body
   by Areas of Higher Elevation
Distance to the Nearest Surface
   Water
Site Under Tidal Influence
Distance to Coastal Wetland
Distance to Fresh Water Wetland
Distance to Critical Habitat
Population Affected Under Air
   Route
Radius Within Which Population
   is Measured
Distance to Commercial/Industrial
   Area
Distance to Park, Forest or
   Wildlife Reserve
Distance to Residential Area
Distance to Agricultural Land

Distance to Prime Agricultural
   Land
Historic Landmark Within View
Site Zip Code
Site Latitude
Site Longitude
Site Ownership
Active Site
Form 2070 File Date
Waste Type
Hazardous Condition
Permit Information
Site Activity
Surface Water Drinking Supply
Soil and Vegetation
Generator on Site
Number of Chemicals Found at
   Each Site
Concentration Level Known
Chemical Observed in Surface
   Water
Chemical Observed in Air
Chemical Used to Score Ground
   Water Toxicity/Persistence
Position of Chemical in Input File
   (i.e., the nth Chemical Listed for
   Facility X)
Chemical Used to Score Ground
   Water Toxicity/Persistence
Chemical Used to Score Air
   Toxicity

Note:

[1] The variable "Physical State" is scored using the same criteria under both
Ground Water Route (GS) and Surface Water Route (SP).

[2] The variable "Toxicity and Persistence" is scored using the same criteria
under both Ground Water Route (GTOX) and Surface Water Route
(STOX) although each route may be scored based on different chemicals.

[3] The variable "Hazardous Waste Quantity" is scored using the same criteria
under Ground Water (GQ), Surface Water (SQ) and Air Route (AQ).

[4] The variable "Distance to a Sensitive Environment" is scored using the
same criteria under both Surface Water (SENV) and Air Route (AENV).

APPENDIX C

Synopsis of Published and Unpublished Studies or Study Protocols Related to Health Effects Evaluations at Hazardous Waste Sites

I. Reference: CDC. Polychlorinated Biphenyls (PCBs) Exposure at Superfund Waste Sites (draft), August 23, 1983 Study Protocol

II. Facility:
   A. Location: Superfund sites with PCB exposures to humans
   B. Physical condition:
   C. Hazardous materials: PCBs and possibly other chemicals
   D. Principal routes of exposures: Various

III. Study:
   A. Motivation of study: There are three reasons given for studying PCBs:
      1. They are the most prevalent chemicals found in Superfund sites.
      2. They can be reliably measured in serum.
      3. Serum levels reflect body burdens over many years due to equilibration between fat PCB stores and serum levels.
   B. Population studied: It is proposed to study people exposed to PCBs from Superfund sites.
   C. Health measurements: Not stated at present.
   D. Exposure measurements: Serum collected from people; questionnaire regarding possible contamination.
   E. Study design:
      Stage 1: Assess the presenee of PCBs at waste sites and possible PCB contamination to humans. The presence of other toxic chemicals will also be assessed.
      Stage 2: A pilot study will be conducted of the potentially "most exposed" people at the high priority sites chosen in Stage 1.
      Stage 3: Community surveys will be conducted at sites found important in Stages 1 and 2.
      Stage 4: Registries of PCB-exposed cohorts will be designed and implemented for cohorts detected by Stage 3.

I. Reference: Halperin, W., Landrigan, P.J., Altman, R., Iaci, A.W., Morse, D.L., and Needham, L.L. Chemical fire at toxic waste disposal plant: Epidemiologic study of exposure to smoke and fire. *J. of Med. Soc. of N.J.* Vol. 78, No. 9, pp. 591–594, August, 1981.

II. Facility:
   A. Location: Southern New Jersey

B. Physical condition: Waste-chemical disposal plant consisting of storage tanks, high-temperature incinerators, ancillary buildings

C. Hazardous materials: Diverse chemicals including PCB, benzene, methylene chloride, aniline

D. Principal routes of exposures: Air, direct contact

III. Study:
  A. Motivation of study: Exposed people developed symptoms. Also there was a possibility that TCDD or TCDF may have been produced in the combustion of PCB.
  B. Population studied: All people present at the fire.
  C. Health measurements: Questionnaires measured previous medical history, complaints of respiratory problems, headache, eye and throat irritation, nausea, etc.
  D. Exposure measurements: Soil samples throughout fire zone and wipe samples of fire equipment measured for TCDD and TCDF. Closeness of people to fire site and duration of exposure were noted.
  E. Study design: Names of people at the fire were obtained by contacting fire companies, police, rescue squads, civil defense groups and other groups that participated, as well as general media appeal. 440 of the 459 people known to have been present at the fire were surveyed. Followup questionnaires were administered to people who reported respiratory symptoms on the first questionnaire.
  F. Limitations of study: No control group of unexposed people, or people exposed to a 'non-toxic' fire. There was no measurement of fumes.
  G. Results: A positive association of symptoms with duration of exposure. No evidence of TCDD or TCDF in any environmental sample.

 I. Reference: Kreiss, K., Zack, M.M., Kimbrough, R.D., Needham, L.L., Smrek, A.L., and Jones, B.T. Cross-sectional study of a community with exceptional exposure to DDT. *JAMA* 245, #19, (1926–1930), 1981.

II. Facility:
  A. Location: Triana, Alabama
  B. Physical condition: Between 1947 and 1971, DDT was manufactured 10 km from Triana. Industrial waste was dumped into a rural stream.
  C. Hazardous materials: DDT and related compounds

D. Principal routes of exposures: Food chain (fish)

III. Study:
   A. Motivation of study: Publicity concerning high DDT residues in fish caught near Triana.
   B. Population studied: All residents of Triana and its rural environs were asked to participate; not all did. 518 people were registered for the study; 499 gave blood.
   C. Health measurements: Demographic, diet, sources of insecticide exposure, medical history, blood analysis, blood pressure, urinalysis.
   D. Exposure measurements: Serum DDT residues.
   E. Study design: Cross-sectional survey of a population likely to be exposed to DDT through food.
   F. Limitations of study: No control group. Complete ascertainment of population at risk was not addressed.
   G. Results: Serum triglyceride was positively correlated with DDT. There was no association with other measured health effects such as history of heart disease, hypertension, diabetes.

I. Reference: Rothenberg, R. Morbidity study at a chemical dump – New York. *MMWR*, Vol. 30, No. 24, pp. 293–294, June 26, 1981.

II. Facility:
   A. Location: Hyde Park landfill, north of Niagara Falls, New York.
   B. Physical condition: It was used from 1953 to 1975 as disposal site for approximately 80,200 tons of chemical waste. A compacted clay cover was installed in 1978 and a drainage system around the perimeter in 1979.
   C. Hazardous materials: Diverse chemicals, including many chlorinated hydrocarbons
   D. Principal routes of exposures: Air

III. Study:
   A. Motivation of study: Not stated
   B. Population studied: People who worked or lived near Hyde Park landfill. There was a 59% participation rate.
   C. Health measurements: Questionnaire (from HANES) and a limited physical examination, urine and blood tests. Compared to matched controls from HANES study.
   D. Exposure measurements: Exposures to people were checked through urine and blood tests. NIOSH had found lindane, mirex and dioxins in dust samples of rafters of nearby companies.

E. Study design: Cross-sectional study.

F. Limitations of study: Low participation rate. Lack of sufficient interval for exposures to produce outcomes such as cancer.

G. Results: Of 180 evaluated variables, 9 were statistically significant; hiatal hernia survey was highest, O.R. = 7.6.

I. Reference: Oudbier, A.J., Eyster, J.T., and Lock, J.M. Berlin-Farro respiratory study, Michigan Department of Public Health, December, 1981.

II. Facility:

A. Location: Genesee County, Michigan

B. Physical condition: Berlin-Farro Liquid Incinerator Company operated from about 1970 to 1974. It was shut down because of resident complaints of fumes and health problems. In 1981 the contaminated sludge was hauled away.

C. Hazardous materials: Diverse chemicals

D. Principal routes of exposures: Air

III. Study:

A. Motivation of study: Resident complaints.

B. Population studied: All residents, 16 years and older, within a 2-square mile around the site.

C. Health measurements: Self-reported respiratory complaints.

D. Exposure measurements: Soil sampling. The exposure measurement of people was self-reported exposure to fumes and length of residence.

E. Study design: Questionnaire survey.

F. Limitations of study: Self-selection and self-reporting bias.

G. Results: Among males, there was an association between respiratory complaints and self-reported exposure, accounting for age and cigarette smoking. Among females there was no association.

I. Reference: New Jersey State Dept. Health. A Health Census of a Community with Groundwater Contamination—Jackson Township, 1980.

II. Facility:

A. Location: Jackson Township landfill, Ocean County, New Jersey

B. Physical condition: Landfill operating from 1972 to 1980. Not insulated from underlying groundwater.

C. Hazardous materials: Chemicals include various halogenated organic compounds.

D. Principal routes of exposures: Groundwater, air.

III. Study:
A. Motivation of study: Demonstrated contaminated well water. A previous health survey showed some complaints. Health department was concerned about long-term effects.

B. Population studied: Affected residents and controls. Total of 560 people from 150 households.

C. Health measurements: Health questionnaire.

D. Exposure measurements: Household water sample. Home air sample. Information on water supply.

E. Study design: Cross-sectional study of exposed and nonexposed controls.

F. Limitations of study: Possibility of recall bias. Residents were aware of problems. There had been previous surveys, and there was a class-action suit in 1980.

G. Results: Increased risk of skin complaints. Among users of shallow wells there was increased kidney illness, but a causal relationship is not biologically plausible.

I. Reference: New Jersey State Dept. of Health. A Health Survey of the Population Living Near Gloucester Environmental Management Services (GEMS) Landfill. April, 1982.

II. Facility:
A. Location: Camden County, New Jersey

B. Physical condition: Landfill was opened in 1960 as sanitary landfill. It is believed that chemical dumping took place in 1970s. It is now covered, but erosion is evident with leachate flowing into an adjacent creek.

C. Hazardous materials: Diverse chemicals

D. Principal routes of exposures: Groundwater, air

III. Study:
A. Motivation of study: Resident complaints of odors and health problems

B. Population studied: Residents living near GEMS landfill and controls living 5 miles away.

C. Health measurements: Questionnaire covering problems like eye and throat irritation, coughing, medical problems. Lung function test.

D. Exposure measurements: Air sampling for volatile organics inside and outside of homes of residents and controls.

E. Study design: Cross-sectional survey of residents and controls.

F. Limitations of study: The initial lung function testing was wrong, so people had to return. Volunteering twice could possibly be source of selection bias. Also, they invited ony some subgroups and not everyone.

G. Results: Increased respiratory complaints in residents, but no medical problems. Pulmonary results were not consistent with questionnaire survey.

I. Reference: New Jersey State Dept. Health. A Health Survey of the Population Living Near the Price Landfill. July, 1983.

II. Facility:

A. Location: Price Landfill, Atlantic County, New Jersey

B. Physical condition: 26-acre site. Sanitary Landfill 1972 to 1980. Closed, inactive, covered, with erosion with leachate on western edge of site.

C. Hazardous materials: Diverse chemicals including cadmium, lead, chloroform, vinyl chloride, benzene

D. Principal routes of exposures: Groundwater

III. Study:

A. Motivation of study: Contamination of private and public wells had already been established by EPA, DEP, and Atlantic County Health Department.

B. Population studied: Households thought to be affected by contaminated groundwater and suitable controls.

C. Health measurements: Health questionnaire on exposure to toxic substances, presence of symptoms and reported medical problems.

D. Exposure measurements: Well water samples showed that Water Quality Criteria was exceeded for many chemicals including cadmium, lead, chloroform, vinyl chloride.

E. Study design: Cross-sectional survey of exposed residents and nonexposed controls.

F. Limitations of study: Bias due to resident knowledge and concern of water problems is unknown.

G. Results: Female exposed residents reported excess of several symptoms compared to controls. There was no difference in reported medical problems.

I. Reference: Janerich, D.T., Burnett, W.S., Feck, G., Hoff, M., Nasca, P., Polednak, A.P., Greenwald, P., and Vianna, N. Cancer incidence in the Love Canal Area, *Science* 212, 1404–1407, 1981.

II. Facility:
  A. Location: Love Canal, New York.
  B. Physical condition: Waste burial 1920 through 1953, population grew rapidly afterwards, though it was affected in 1978 when concern started regarding health hazards. Inactive landfill in residential area.
  C. Hazardous materials: Over 80 chemicals identified (though not all toxic), including benzene and halogenated hydrocarbons. Largely residues from pesticide production.
  D. Principal routes of exposures: Direct, air, water.

III. Study:
  A. Motivation of study: This was one of later studies of a recognized problem area.
  B. Population studied: Exposed: Exposed people taken from census tract surrounding Love Canal (LC). Unexposed: Unexposed people taken from Niagara Falls census tracts and New York state rates. Estimated from 1960, 1967, and 1970 censuses.
  C. Health measurements: All sites cancer incidence as reported to the New York Cancer Registry.
  D. Exposure measurements: None in this study.
  E. Study design: Ecologic design. Age/sex-specific cancer rates of LC area were first compared to 25 Niagara Falls census tracts, then New York state incidence rates. All sites compared for 1966 to 1977 (better data). Liver, lymphoma, leukemia compared for 1955 to 1965 as well.
  F. Limitations of study: Population at risk ill-defined. Completeness and accuracy of registry data not known; it could be a factor since there were only a few deaths per census tract per age/sex category in several cancers.
  G. Results: There was high rate of respiratory cancer, but it seemed to be related to entire city of Niagara Falls. Standardized incidence ratios based on New York state rates found the 2 female 1955 to 1965 liver cancers significant. Another look at age-specific rates comparing LC to Niagara Falls census tract found LC to be high. But the authors discounted all positive findings because of the small number of cases.

I. Reference: Indian et al. Health Survey of a Population in the

Proximity of a Chemical Waste Disposal Site, Lorain County, Ohio Department of Health. December, 1980.

II. Facility:
   A. Location: Lorain County, Ohio
   B. Physical condition: Chemical disposal site, begun in 1950s. Began with open burning of chemical wastes but has since used a series of improved incinerators. Clay lined landfills and lagoons are utilized.
   C. Hazardous materials: Diverse chemicals.
   D. Principal routes of exposures: Air.

III. Study:
   A. Motivation of study: Request by EPA to investigate resident complaints.
   B. Population studied: Northern Eaton township residents were target population.
   C. Health measurements: Cancer mortality and interview survey on health complaints.
   D. Exposure measurements: None, but there was an interview survey of nuisance complaints.
   E. Study design: Ecologic design. In the part of the study discussed here, death certificates were obtained from the Ohio Department of Health. Observed cancer mortality for study area was compared to expected based on mortality from control populations. Also, house-to-house survey done on health complaints.
   F. Limitations of study: The rural control populations differed by socioeconomic factors in the mortality study. Lifestyle factors were not considered. Small populations meant small statistical power to detect differences. Survey results were not compared to any controls.
   G. Results: No significant increase of mortality. Residents complained of odors from waste site.

   I. Reference: Lyon et al. Cancer clustering around Point Sources of Pollution: Assessment by a Case-Control Methodology. *Env. Res.* 25, 29–34, 1981.

  II. Facility:
   A. Location: Utah
   B. Physical condition: Uranium tailing dump
   C. Hazardous materials: Radon gas
   D. Principal routes of exposures: Air

III. Study:
  A. Motivation of study:
  B. Population studied: People living in the two counties containing dump site between 1966 and 1975.
  C. Health measurements: None—just what was reported to cancer registry.
  D. Exposure measurements: Environmental measurements were taken around dump, but not for individuals.
  E. Study design: Ambidirectional design. Cases and controls were drawn from Utah Cancer Regstry, their addresses determined. Distance from site was used as substitute exposure measurement.
  F. Limitations of study: Distance from site may not be a good exposure measurement. It does not account for wind direction. Assumption that migration and ascertainment for cases and controls be equal may not be valid. An excess of lung cancer in males might be due to men living close to where they work.
  G. Results: No significant differences between cases and controls for radiogenic malignancies (acute leukemia, multiple myeloma, osteogenic sarcoma).

  I. References: Clark et al. An Environmental Health Survey of Drinking Water Contamination by Leachate from a Pesticide Waste Dump in Hardeman County, Tennessee. *Arch. Environ. Health* Vol. 37, 9–18, 1982.
  Meyer, C.R. Liver Dysfunction in Residents Exposed to Leachate from a Toxic Waste Dump. *Environ. Health Pers.* 48, 9–13, 1983.

 II. Facility:
  A. Location: Hardeman County, Tennessee
  B. Physical condition: Land dump operated from 1964 to 1972 by local pesticide manufacturer. 200 acres. 3,000,000–500,000 55-gallon barrels of liquid and solid waste in shallow trenches. Closed in 1972 because of contaminated water in test wells.
  C. Hazardous materials: Diverse, including chlorinated organic compounds.
  D. Principal routes of exposures: Ground water, air.

III. Study:
  A. Motivation of study: In-progress study of Memphis wastewater treatment plant workers who treated contaminated water plus complaints of local citizens of skin and eye irritation, muscle weakness, shortness of breath, nausea, vomiting, diarrhea, abdominal pain.

B. Population studied: Hardeman County resident wells were tested. People were divided into exposed, intermediate, control. It is not clear who was eligible for study, how many were tested, etc.

C. Health measurements: Medical Questionnaire (alcohol/drug consumption, history of liver problems), blood (liver profile), bile acid survey, urine analysis.

D. Exposure measurements: Air and water samples from some homes.

E. Study design: In November 1978 a survey of 36 exposed individuals was conducted. In January 1979 another survey was done; people were categorized by exposure: 49 were exposed, 33 were intermediate-exposed, 57 were nonexposed. 31 people participated in both studies. It was a hasty study because by end of 1978 people were not using wells (being closed).

F. Limitations of study: Population at risk and controls ill-defined. Study initiated by complaints of residents (bias).

G. Results: Apparently, the November 1978 exposed were compared to the January 1979 nonexposed to find significant differences in liver enzyme profiles (alkaline phosphatase, SGOT). However, there was a decrease of liver enzyme values in the 31 people measured in November 1978 and January 1979. So, January 1979 study found nothing significant. Result: a subclinical transitory liver insult. No differences in skin or eye abnormalities.

I. Reference: Parker, G.S., and Rosen, S.L. Woburn: Cancer Incidence and Environmental Health Hazards, 1969–1978. Massachusetts Department of Public Health. January 23, 1981.

II. Facility:
  A. Location: Woburn, Massachusetts
  B. Physical conditions: In 1979 industrial wastes were discovered in a section of Woburn called Industri-plex. Two drinking water wells were closed.
  C. Hazardous materials: Chemicals include lead, arsenic, organic contaminants.
  D. Principal routes of exposure: Groundwater.

III. Study:
  A. Motivation of study: Detection of wastes and suspected high incidence of cancer.
  B. Population studies: Residents of Woburn
  C. Health measurements: Vital statistics on childhood leukemia, renal, liver and bladder cancer.
  D. Exposure measurements: Possible exposure to known carcinogens

were obtained through interviews. Several wells were tested for arsenic, chromium, and lead.

E. Study design: Ambidirectional design. Cases of cancer of residents were ascertained for the years 1969 to 1978. Two matched controls were chosen for each case. Cases and controls (or their nearest relatives) were interviewed.

F. Limitations of study: There were small numbers: twelve cases of childhood leukemia, thirty cases of renal cancer.

G. Results: Although the leukemia and renal cancer rates were higher than expected by comparison to data from the Third National Cancer Survey, the case-control interviews revealed no association with environmental exposure.

I. Reference: Pennsylvania Department of Health. Health Surveillance Protocol: Planned Health Activities Related to the Drake Chemical Superfund Site, Lock Haven, Clinton County, Pennsylvania. November 15, 1983.

II. Facility:
  A. Location: Clinton County, Pennsylvania
  B. Physical condition: Kilsdonk Chemical Corporation operated from 1948 to 1962. Drake Chemical Company continued operations until 1981, when it went bankrupt. There is concern about chemical storage problems at the site.
  C. Hazardous materials: Diverse chemicals, including beta-naphthylamine, DCB, TCPAA.
  D. Principal routes of exposure: Air, direct contact

III. Study:
  A. Motivation of study: Previous health studies revealed adverse health effects related to the site. Request for Pennsylvania Department of Health involvement came from EPA, DER, and local citizens.
  B. Population studied: There will be different populations associated with the different substudies.
    1. Clinton County residents will be studied in the mortality survey.
    2. Residential population at risk (retrospective) will be studied in morbidity study.
    3. Occupational study will cover employees at Drake who worked at least 3 months from 1940 to 1981.
  C. Health measurements:
    1. Cancer mortality

2. Mortality and morbidity ascertained through survey questionnaires and followup of hospital/medical records
3. Medical surveillance includes urinalysis
D. Exposure measurements: Unspecified determination of environmental and occupational exposure.
E. Study design:
1. Ecologic examination of cancer mortality rates 1950 to present
2. Cross-sectional survey of residents
3. Medical surveillance of former workers

I. Reference: Talbott, E., Radford, T., Schmeltz, R., Murphy, P., Kuller, L., Portocarrero, C., and Doll, R. Distribution of thyroid abnormalities in a community exposed to low levels of gamma radiation. Submitted to *Amer. J. Epidemiol.*, 1983.

II. Facility:
A. Location: Strabane/Canonsburg, Pennsylvania
B. Physical condition: A refinery processing uranium began its operation in 1910. Radioactive waste continued to be dumped around the plant until 1958. A survey in 1977 showed that onsite levels of gamma radiation were high (500 R/hr). Liquid wastes were discharged into an open drain which emptied into a swamp.
C. Hazardous materials: Radioactive waste.
D. Principal routes of exposures: Air, direct contact.

III. Study:
A. Motivation of study: A DOE report of onsite and offsite gamma radiation levels became public. Graduate School of Public Health, University of Pittsburgh initiated a study.
B. Population studied: Exposed: current full-time resident, at least 21 years of age who lives within 1/3 miles of site and has been a resident in the area for at least 15 years. 514 people were eligible for the study. Unexposed: residents of a nearby community, Muse, Pennsylvania, who are at least 21 years of age and have lived in the vicinity for at least 15 years.
C. Health measurements: Thyroid evaluations, medical history, demo.
D. Exposure measurements: None. Length of residence was the indirect measure of exposure.
E. Study design: Cross-sectional study of prevalence of thyroid disease in an exposed population and unexposed population matched for several demographic variables.
F. Limitations of study: Definition of outcome: defining which thyroid diseases are rad-related is difficult. There may be a

self-selection bias in the decision to participate; thyroid disease may be more frequent among participants than nonparticipants.

G. Results: Rate of radiation-related thyroid diseases in exposed population was marginally higher than in the nonexposed population.

I. Reference: Paigen and Goldman. Lessons from Love Canal. The Role of the Public and the Use of Birthweight, Growth and Indigenous Wildlife to Evaluate Health Risk. 4th Annual Symposium on Environmental Epidemiology.

II. Facility:
   A. Location: Love Canal, New York
   B. Physical condition: Residential area over and near capped toxic waste dump
   C. Hazardous materials: Diverse chemicals, residues from pesticide production.
   D. Principal routes of exposures: Direct, air, water

III. Study:
   A. Motivation of study: Concern of local residents. Residents received grant money and initiated a study not involving New York state or Feds.
   B. Population studied: 523 children, 18 months to 16 years living in Love Canal area; 440 controls.
   C. Health measurements: Birth weight, birth height
   D. Exposure measurements: None. The only exposure-related measurement was distance from dump site.
   E. Study design: Ambidirectional design. Comparison of birth weights and birth heights of exposed and unexposed residents and controls.
   F. Limitations of study: The possibilities of selection or misclassification biases are discussed in the chapter. Regarding misclassification bias, there is no certainty whether those classified as exposed really had exposure, or whether those classified as unexposed really were unexposed.
   G. Results: Odds ratio for low birth weight children among Love Canal homeowners was 3.0. Children living in Love Canal were also shorter than the controls.

APPENDIX D

OSHA and Risk Assessment: A Review of Public Comments

1. Executive Order 12291. (2/19/81)

    (a) Industry commenters remarked that in effect, Executive Order 12291 requires that setting priorities for regulations "to the extent permitted by law," includes a cost-benefit analysis. (These cost-benefit analyses would be included in mandated regulatory impact analyses.)

    (b) The general tenor of the public-organized labor comments were that OSHA ". . . is permitted to utilize cost/benefit analysis in setting priorities, and Executive Order 12291 sets the analytic framework for analysis . . ."

    At least one (industry) commenter asserted that in the cotton dust case the Supreme Court rejected cost/benefit analysis as a primary rule-making consideration, and that any "prioritizing process must give first consideration to the health and medical aspects."

    Several other commenters (from industry and organized labor) separated the identification of carcinogens (hazards) from their management, arguing that the Cancer Policy should concentrate on scientific issues.

    Only one commenter raised questions about the adequacy of the data to support OSHA's carcinogen classification scheme.

Total number of commenters	28
Industry (including legal representatives)	23
Public, organized labor, other government	5

2. Cost-benefit analysis (in general)

(a) Industry commenters usually agreed with this statement: "Cost-benefit studies should be an integral part of all decision making, and especially those which deal with the health of humans."

(b) Labor and public commenters found that "business grossly over-estimated the expense involved. Economic costs are recoverable. Loss of health or life is not."

Total number of Commenters	11
Industry	7
Public, labor	4

3. Should techniques of quantitative risk assessments and significant risk determination be specified?

(a) Major industry comments:

1. "The art of risk assessment still is in its infancy . . . Specification of a procedure . . . would close the door to the use of any future development."

2. "An independent group of assessment experts should be formed to make an assessment of . . . each chemical's use."

3. "It is possible and perhaps desirable that a policy should specify methods."

4. OSHA should review the state of the art and provide nonmandatory guidelines. These guidelines should be independent of the Cancer Policy.

(b) Public and labor comments included:

1. ". . .techniques [are] too detailed to specify."

2. Risk assessment is almost impossible to use in some industries — such as construction.

3. ". . . risk assessment technology — is still at an early . . . stage of development."

4. "The policy should set the standard — not the details."

(c) Three "public" commenters (scientists who had published on the problem) noted the paucity of exposure data from which to make risk assessments.

Total number of commenters	35
Industry	25
Public, labor	10

4. Choice of proposed quantitative risk assessment models or methods

   (a) Industry comments:

   1. "We do know that the linear-through-zero methods have been found inappropriate in every case where adequate animal/human data have been obtained."

   2. "Negatives should be used." (as well as "positives")

   3. No single mathematical model for low dose extrapolation is recommended. Models should be selected on the basis of known biological mechanisms. Use several models and report a range of results.

   4. The lower 95% confidence limit on the dose giving 10^{-2} risk is a permissible level.

   (b) Public, labor comments:

   1. "Any quantitative risk assessment must consider [a whole range of] biological processes ranging from genetic variability to aging, and including interactions.

   2. "OSHA should be requesting information which would show that exposure estimates are sufficient for quantitative risk assessment to be undertaken."

Total number of commenters	10
Industry	8
Public, labor	2

5. Proposed criteria for determining the significance of risk

   (a) Industry comments:

   1. A positive result in one test at one (high) dose is not sufficient support for a finding of significant risk in humans—Compare health risks to other risks—Define an "insignificant risk "as a threshold—Consider economics, reduction in consumer choice, etc.

2. Use Todhunter criteria—i.e., $(1 \times 10^{-5}$ or 1×10^{-4} lifetime risk). "To use risk estimates as absolutes is inappropriate." Consider metabolic pathways.

3. Compare carcinogenic risk to other industrial hazards. Make relative risk evaluation. "The Policy should not contain rigid criteria for what is a significant risk."

4. "Guidelines will, of necessity, have to be very general."

(b) Public, labor comment:

1. "The original cancer policy is itself a well-researched significant risk assessment."

Total number of commenters	13
Industry	12
Public, labor	1

6. Overall effects of the policy (including legality of policy)

(a) Industry comments:

1. Standards are unnecessarily rigid, expensive to meet and not based on adequate data. "The opportunity to achieve much more substantial protection in other areas at lower cost is missed."

2. The policy (appropriately amended) should not impose any costs on industry or the economy as a whole.

3. The current policy is in violation of the OSHA act and Supreme Court decisions. OSHA must show "significant risk" and that the standard will bring about a significant reduction in risk.

4. The use of generic rules is contrary to the statutory requirement for the use of the best available data.

(b) Public, labor comments:

1. "The Cancer Policy at present is in conformity with recent Supreme Court decisions . . . The Court indicated that the amendments made in the Cancer policy in light of the Benzene decision . . . had been responsive to the Court's interpretation of the OSHA act."

2. "In January 1981 OSHA promulgated deletions to the Cancer Policy to remove any provisions which contradicted the Supreme Court's benzene decision."

Total number of commenters	20
Industry	16
Public, labor	4

Index